SOTHIS PRESS

Sothis Press aims to completely reimagine the public domain book with an emphasis on universal access to captivatingly designed print editions. We believe that technology will enable every single book ever published to be perpetually accessible in beautiful print form. Sothis is beauty in ubiquity.

www.sothis.press
1 Waterhouse Square, London, ECIN 2ST

Practical Road Building

Endorsed by the National Highways Association
and the American Automobile Association

Charles E. Foote

SOTHIS PRESS

ISBN 978-1-0368-0933-1

PRACTICAL ROAD BUILDING

BY

CHARLES E. FOOTE

ENDORSED BY

The National Highways Association

AND

The American Automobile Association

PHILADELPHIA

DAVID McKAY, Publisher

604-608 S. Washington Square

CONTENTS

PART I

CHAPTER I

CHAPTER II

CHAPTER III

ILLUSTRATIONS

PREFACE

THE call for a book on "Practical Road Building" reached the author from a variety of different sources during the past three or four years. The most impressive of these came from personal experiences when delivering lectures in various sections of the country on "Highway Construction," "Highway Bonds," "Road Materials," and other phases of the Road Industry, when farmers, bankers, merchants, local officials, local capitalists, local producers, and local vendors submitted questions readily answerable except for the limit of time.

"Where can the information be obtained in concrete and intelligible form?" was the final question. And as it could not be answered, this volume is intended to meet the requirements.

The Federal Office of Public Roads and Rural Engineering readily supplies information concerning any detail. A number of technical works on Highway Engineering are of the highest excellence; a multitude of papers and addresses, read and delivered at various

Highway and Engineering Conventions, and the reports of various committees of technical societies, all contain information of great value, but which is unavailable to the lay reader—the average local official and the citizen—because of their technical character and form of expression.

In this volume the fullest possible use has been made of all such technical information as would fit the conditions, and the author desires to express his fullest appreciation and to extend the greatest possible credit to those from whose works he has abstracted information and turned it into practical and intelligible language for non-technical readers. To give the names of those whose writings have been in part appropriated would be to prepare a list of all the men prominent in the Road Industry during the last ten years; a list too long for these pages.

The fact that the great State and Federal appropriations for improved roads will apply mostly to main roads, which will be in charge of presumably competent highway engineers, does not alter the fact that fully 80 per cent. of the roads of the country will for many years remain in the hands of local officials and be subject to the influences of local necessities and local sentiment. While many engineers may gain knowledge and information from these pages by careful study, the primary purpose of the author has been to present a work which will give full information to the reader who wants it presented in plain language without technicalities.

The closing months of 1916 emphasized the necessity for road improvement in the United States more forcibly than any other period during two generations of men. The cost of primary marketing; the difficulties of prompt transportation, and the excessive prices at points of consumption have made a community of interest between the producer and ultimate consumer which should go far to show their interdependence on each other, and their mutual interest in the reduction of primary transportation expense by means of improved highways.

The destinies of the Nation depend on the conservation of its resources. In no one direction in the past has there been more waste than that caused by bad roads. If the information presented in these pages shall to any considerable extent add to the intelligent extension of country road building with the consequent advantages to the communities, States, and Nation, the author will consider that the work involved in its preparation has been of value to the people of the country.

CHAS. E. FOOTE.

NEW YORK,
January 1, 1917.

FOREWORD

ROADS AND THE ROAD USER

WITH over 3,500,000 automobiles in use in the United States, or an average of 1 for every 29 people, it is only natural that there should be the widest interest among motorists in the improvement, construction, and maintenance of public highways. It is altogether probable that the ultimate number of automobiles will be double the number now in use.

It is estimated that the investment in automobiles now is $3,000,000,000, and that the ultimate investment will be $6,000,000,000. It has been estimated that the average distance that each automobile is driven yearly is 4000 miles, which would make the staggering total of over 14,000,000,000 car miles every twelve months, with an ultimate of probably 28,000,000,000 car miles. The average annual cost of operating an automobile is $250, which would make a total annual operating charge of $878,000,000, or an ultimate of $1,757,000,000.

In view of these figures, it is not astonishing that the average motorist should appreciate the necessity of larger expenditures than ever before for road improvements.

Road improvement is as necessary, from the standpoint of the motorist, as gasoline, oil, tires, and car repairs. A certain part of such cost must be assumed in computing automobile expense. Improper construction and inadequate maintenance are not only a source of annoyance and inconvenience to the motorist, but are just as much a source of expense as a poor quality of gasoline or improperly built tires.

The motorist constantly passes over city and township lines, goes from one municipal subdivision to another, so he desires smoothly paved city and village streets as well as improved country roads. The preponderating percentage of good roads are built—

In Cities, through their departments of public works,

In Parks, by the commissioners in charge,

In Villages, by local street commissioners,

In Townships, by township boards and highway commissioners,

In Counties, by county commissioners, or supervisors,

In States, through state highway commissioners,

In the Nation, pursuant to the recently enacted Federal Aid Road Act.

The improvement of city streets and park boulevards are of ever-increasing importance, yet the greatest mileage of through-roads must be constructed in the rural localities in whole or in part by the four political subdivisions, viz.: the Town, County, State, and Nation.

The permanent roads outside of cities and villages in New York State, upon the completion of the proposed systems, will represent a capital investment of $300,000,000, or about equal to the amount invested in automobiles. There are 80,000 miles of highway in that commonwealth, so the average capital investment is about $3750 per mile of highway.

There are approximately 2,000,000 miles of highway in the United States. At $3750 per mile, this would represent a capital investment of $7,500,000,000. The ultimate value of automobiles will be upward of $6,000,000,000.

The maintenance, reconstruction, and upkeep will represent possibly 5 per cent. of the capital on $375,000,000, about one-quarter of the ultimate automobile operating charge.

The preparation of road plans, the selection of types of construction, and the various details must be determined by competent highway engineers, appointed on merit and assured of a reasonable tenure in office.

All of the several types of road described in this

volume have their place in one or another of the classes of improvement. There is no type of road which is applicable to all classes, and any work which adds to the popular knowledge of this subject is of value. It is with this purpose that the American Automobile Association endorses this volume, which presents many interesting road matters in a form to be appreciated by the average motorist.

GEORGE C. DIEHL,
Chairman A. A. A. Good Roads Board.

WASHINGTON, D. C.,
May, 1917.

FOREWORD

The title of this book, "Practical Road Building," carries with it the author's purpose. Not a technical treatise for engineers or scientists, and yet an engineering book for the road builder who is called upon to build and who must build, and at once, with but little if any scientific education himself or the means of its employment. And yet also a book full of valuable practical information for the engineer himself, and stated in common-sense language devoid of hair-splitting, technical nomenclature.

There are many thousands of road officials in our small towns, villages, townships, and counties who will find it easy of reading and understanding and valuable in its shortness, simplicity, and directness of statement and purpose. It appeals more directly to the practical lay mind, and yet has great value to the average engineer as well. The author, Mr. Charles E. Foote, has for many years devoted himself to the practical side of road improvement. He is well qualified by his knowledge and experience as a writer to set forth the subject from its practical viewpoint.

In appreciation of the wide general value of such treatment of the subject the National Highways Association has endorsed this volume as the first of a series. Others on "Road Machinery" and "Highway Materials" will follow, largely through the interest and co-operation of David McKay, the publisher, whose exceptional facilities for their distribution should give this volume, and those following, a large National audience. This is the hope and wish of the National Highways Association.

CHARLES HENRY DAVIS, C. E.,
President National Highways Association.

PRACTICAL ROAD BUILDING

PART I

CHAPTER I

A SKETCH OF ROAD HISTORY

THE most insistent demand made by civilization at the beginning of the twentieth century is for better means of communication; that communities and peoples may keep in closer touch with one another; that commerce may become more and more liquefied; that social and moral welfare may be enhanced, and that the people, by reason of this closer contact with each other, may keep up with the spirit of the age in culture and righteousness and wealth. To accomplish these ends requires the improvement of the public highways.

From the beginning of civilization history makes frequent mention of the building of roads, and in some instances modern excavations have disclosed the existence of great highways which the historians have overlooked. As nearly as may be ascertained those ancient highways were built for military purposes; for the more rapid movement of armies and the transportation of munitions of war; but they served the purpose, also, of channels of commercial and social communication and in this manner aided in the world's intellectual development.

Nearly four thousand years ago the city of Ancient Babylon, the city made famous by the biblical story of Belshazzar and the handwriting on the wall, built a road to Nineveh, a distance of nearly 400 miles. The road was broad and well graded, and was paved with brick set in a mortar of asphalt. Sections of this road are still found, buried under the accumulations of centuries, when the tools of the archæologist bring them once more to the light of day.

In Egypt, near the Great Pyramids of Gizeh, have been found the ruins of a great highway, or causeway, more than a mile long. Herodotus mentions that the great King Cheops built such a road for the transportation of materials used in building the Pyramids. He states that the road required the work of 100,000 men for a period of ten years. It is probable that the ruins recently discovered are a part of the same highway. It was built of stone blocks some of which are 10 feet thick. Temples, parks, statues, and mausoleums were placed along its sides.

The Persians had a system of military highways which extended for long distances throughout that country. They maintained a post or messenger service, with stations 18 to 25 miles apart, where messengers would exchange their tired horses for fresh ones. In this way a speed often reaching more than 100 miles a day was secured.

According to the historian Strabo, there were two great roads extending between Syria and Babylon. One of these was much more popular than the other because of the fact that less toll was charged.

In ancient Greece much attention was devoted to the subject of highways by the Senate at Athens and the authorities of Thebes and Lacedæmon. Highways were maintained to the Piræus and to the sacred shrines.

An interesting story comes down in the history of Thebes. It appears that Epaminondas, one of the greatest of Theban generals, had failed to capture Corinth; and as a result had been visited with the indignation and contempt of the populace. In order to express more completely their sentiments toward him they elected him to the office of "cleaner of streets," which office at that period was the lowest and most despised of any occupation in that city.

As the story goes, Epaminondas accepted the situation gracefully, and turned his attention to cleaning the streets of Thebes. By virtue of his great ability and his trained mind he made Thebes the cleanest and most beautifully kept city in the world, until its cleanliness became proverbial, and was noted in the literature of the times. "As clean as Thebes" is still a proverb noted in the early Greek classics. Then the office of Cleaner of Streets became the highest office within the gift of the people, and citizens of the highest standing and culture aspired to it.

Most historians, in commenting on the great Roman roads, intimate that the art of road building was learned by the Romans during the long series of wars with Carthage. It is known that the Carthagenians had a great system of highways, which, together with their communication by sea, enabled them to withstand the alternate assaults of Greece and Rome for several centuries. For nearly four hundred years Carthage sustained long and bloody wars. It fell about 146 B. C., and about all that can be found today of its ancient civilization is the ruins of some most excellently built highways.

In its conquest of the world Rome built twenty-nine great highways radiating from the Eternal City into the various provinces. It seems to have been the policy, as soon as a section or country was conquered, to connect it with Rome by a great highway, over which troops could be moved with celerity and over which the tribute, which Rome always demanded, could be carried to the Capital.

One of the most famous of these roads is still known as the "Appian Way." It was begun by Appius Claudius about 312 B. C. and was first built to Capua, about 142 miles. Later it was continued to Brundisium—now Brindisi—making its total length approximately 360 miles. It is supposed to have been completed during the reign of Julius Cæsar. The Roman roads were of very heavy construction, and similar roads today in the United States would cost approximately $245,000 per mile.

We learned in our early Latin studies that when Cæsar started out with his legions for the conquest of Gaul he "crossed the Rubicon" and "burned his bridges behind him." These facts have been paraded to an admiring world for twenty centuries as the highest possible examples of patriotism and determination. They have carried the idea that there was no possibility of turning back, and that the means of retreat had been destroyed, that such a temptation should not occur.

Yet, when we study the progress of that campaign of conquest through France and Belgium and into Britain the fact is borne in on us that Cæsar must have been making what would now be called a "grand-stand play," or perhaps "playing to the galleries," because, as a matter of fact, Cæsar had among his forces a set of the most expert bridge builders in the then known world.

Specimens of their handiwork are still extant in structures of masonry which are still doing service on the highways of France and Britain, where great roads were built by the invading Roman hosts.

It seems very probable that none of the ancients really appreciated the commercial value of their highways. The periods were periods of wars and military operations, and the transportation of troops and supplies was the paramount necessity. This aspect is confirmed by the fact that after the fall of Rome, and during that period of history known as the "Dark Ages" extending down to the ninth or tenth century, when practically all Europe relapsed into feudalism, thousands of miles of highways were destroyed and bridges torn down, because of the fact that they would furnish easier access by enemies. Such civilization as there was seems to have found its expression in architecture, in the construction of great castles, and in making them impregnable. Commerce on land, such as there was of it, was mostly carried on on the backs of horses.

By reason of these conditions the association of peoples with each other became restricted. Education languished and in many regions disappeared, except among the religious orders and in some of the most highly cultivated families. Many kings and princes of those times could neither read nor write.

Road building, that is, the building of good roads, did not soon begin after the human mind had begun to throw off the shackles of isolation, and the people again began to form combinations for mutual advantage, which combinations finally resolved themselves into more or less stable governments. Roads were just what the people of localities made them. In the seventeenth century England passed a law authorizing the seizure of land along a highway for the use of travelers when the road became so deep as to be impassable. Contemporaneous history states that the roads of England at that day were worn into great trenches, sometimes 4 feet deep, in which the water settled after rains. At about the same period laws were enacted requiring that the forests be cleared from each side of the road to a distance of 200 feet, so as to prevent surprises to travelers by the highwaymen and brigands who at that time had become a menace to travel.

The building of improved highways, as the term is understood today, began about 150 years ago in France. In 1764 a provincial engineer named Tresaguet, of Limoges, who had made a local reputation by building roads of broken stone, was called to Paris by King Louis XIV to explain his system. He was then appointed by the king to the position of Assistant Inspector General of the Department of Public Works, in charge of roads and bridges, and began the construction of a system of King's Highways in France.

In the year 1776, when the American colonies were declaring their independence, France adopted as a National Highway System the roads previously built, and those to be built between the city of Paris and the principal cities of the various provinces or departments of the country. The work of construction continued until about 1790, when 15,000 miles of stone roads had been completed.

For probably one hundred years before the appearance of Tresaguet spasmodic efforts had been made at improving the King's Highways. Several prime ministers, including Richelieu, the "Great Cardinal," had made efforts in that direction. Most of the labor, both before and during the administration of Tresaguet, was forced labor, levied upon the peasantry. Some historians consider this one of the contributing causes of the French Revolution, known as the bloodiest tragedy of history.

Studying the question from a broader and deeper and more comprehensive standpoint, it seems that it was the facility of communication offered by these highways that made the Revolution possible. For hundreds of years the peasantry of France had been ground under the heel of despotism. By reason of the isolation of the various sections from each other there could be no common thought or expression. The people were sent to work on the roads or drafted into the army at the pleasure of their royal masters.

While the great highways were built for the primary purpose of moving troops and supplies, they also afforded the means of communication between the people of the different parts of the kingdom, and permitted the development of that irresistible sentiment which caused the population to rise and move, like a steadily growing avalanche, on to the capital where the great tragedy culminated.

Later, Napoleon Bonaparte extended the system of highways, and had cross-roads built. He also put in operation a system of maintaining the roads by putting a patrolman on every 5 or 6 miles, and placing materials along the roadside, so that any defect in the road surface could be treated without delay. This system still prevails, and has been incorporated into the highway systems of most of the countries of Europe and into those of several States of the Union.

The elaboration of the Road System of France by Napoleon, primarily for war purposes, had effects which may not have been contemplated by him. With the National Highways, the Departmental Highways, and the Communal Highways surfaced with stone, the peasantry became rich—the richest class per capita of any farming people on the face of the earth. Education and knowledge spread, and the people were brought into closer communication with each other.

Then came the downfall of monarchy in France. Is it not fair to assume that the downfall was due to the roads which permitted the free interchange of thought throughout the country? That the highways which had been designed as an aid to national defense and offense became also the channels to material wealth, and to that higher intelligence which makes absolutism in government impossible?

French roads are today considered the best in the world, and its systems are the patterns for the road builders of all countries. But the governmental principles under which they were designed have been found wanting in the inexorable evolution of time.

English roads continued in very bad condition from earliest historical periods until the appearance of Telford in the South of England and in Wales, and McAdam in the North of England, about 1815. Telford's plan was to take field stones, set them on edge with the pointed end up, then break off the points with a sledge, and drive the pieces down as wedges to hold the stone firmly. On top of this was placed a layer of broken stone for a wearing surface. This class of foundation is still much used in making road foundations, especially where the ground is low or marshy.

McAdam's plan was the original form of what is still

known as the macadam road. It was made of broken stone, which the traffic would pack down. When the surfaces of the stones would crumble or be broken off into small particles or into dust, the rains would wash the smaller particles into the interstices, and further traffic would hold them there, and thus a smooth, even, hard surface was produced. This method was practically identical with that of Tresaguet of fifty years previous.

Early efforts to build roads in the United States met with little encouragement. The first road law enacted in the colonies, so far as can be ascertained, was by Virginia in 1632. The "New England Path," between Boston and Plymouth, was begun in 1639. The Province of New York enacted laws regulating the building of roads in 1664, and in 1666 Maryland passed a road law. In 1692 Pennsylvania enacted a law placing the control of the roads in the hands of township officials. This was changed to county control in 1700.

But little seems to have been done with the roads for nearly a century, or until after the close of the Revolutionary War. The larger cities were all at the seaboard, and most of the cultivated farms were along or near water-courses which could be used for such transportation of agricultural products as was found necessary.

Within a few years after the independence of the

colonies, and the formation of stable government for the union of States, the necessity for highways became insistent. In 1786 Virginia authorized the construction of a highway across the Cumberland Mountains into Kentucky for the accommodation of the trade of those who had settled and were settling that territory; and when Kentucky became a State in 1792 one of the very early acts of its legislature was to provide for the continuation of the highway in the new commonwealth.

In 1794 was started what is supposed to have been the first broken stone road in the United States—the Philadelphia and Lancaster Turnpike, in Pennsylvania. It was built as a toll road by private capital, and to this day most if not all of the road still collects toll from those who pass over it.

The city of Lancaster, Pennsylvania, has a population of about 35,000 people, and is located in what is claimed to be the richest agricultural county in the United States. But there is no highway by which that city can be entered or departed from without paying toll. Notwithstanding this fact, in a recent election for a state road system the people of the city and county gave a majority against good roads of over 11,000 votes out of a total vote of less than 16,000. In the adjacent county of York a similar condition prevails.

Toll roads, which were common in the early history of the country, have almost vanished, and a few more years will see their total elimination. They served their purpose and served it well. Now, like obsolete factors in human progress, they must give way.

The United States Government went into the road building business in 1806, when what was known as the National Road was authorized by Congress and construction commenced. The original design was to build the highway from tidewater on the Atlantic Coast to a point on the Mississippi River opposite St. Louis, what is now East St. Louis, Illinois. As a matter of fact the construction began at Cumberland, Maryland, which was practically the head of navigation on the Potomac River. The road was built across the slender part of the state of Maryland, through the southwestern portion of Pennsylvania, and across the panhandle of the state of Virginia—now West Virginia—to the Ohio River.

Crossing the Ohio River by ferry, the road took almost a direct line across the states of Ohio and Indiana, passing through the capitals, Columbus and Indianapolis, thence deflecting a little to the southwest, reached across the state of Illinois.

Appropriations were made by Congress from time to time as the work of construction progressed. The engineering work was of the highest order. The road was well graded and surfaced with broken stone or gravel, mostly under the direct supervision of government engineers. The road was completed to the Ohio River in 1817, and furnished the main thoroughfare between the East and the rapidly growing West. Within about twenty years afterward the road had been graded throughout its entire length, and the heavy stone surface and the bridges of stone masonry had been completed as far as Indianapolis. The last congressional appropriation was made in 1838.

As illustrating the short-sightedness of many of the human family, it is worthy of note that when the National Road was proposed, so that stage-coaches and freight wagons could be hauled long distances with comparative ease, the most violent opposition was encountered. It was held by those opposing the road that the then great industry of conveying passengers and freight across the mountains from the Potomac to the Ohio on horseback would be destroyed, and that thousands of men would be out of work, and hundreds of thousands of dollars' worth of animals and equipment be rendered worthless.

Again, when the railroads began to be built from the east to the west, there was the same violent opposition from those individuals and statesmen who could see only the ruin of the stage-coach and team-freighting industry. True, those industries were ruined so far as main line traffic was concerned, but the establishment of better and cheaper means of through transportation not only made opportunities for "feeders" which required all the existing equipment, but many times more.

While the National Road had been building Ohio and Indiana had built thousands of miles of gravel and other kinds of improved roads, most of which led to the National Road, though some connected with the Ohio River on the south, or with Lake Erie on the north.

After the beginning of railroad building the National Road was virtually abandoned. First, Congress turned the various sections of it over to the states in which they were located, provided the states would keep them in repair. Then the states turned sections over to the counties under similar conditions. Then some counties established toll-gates to get money for maintenance; others turned their sections of the road over to turnpike companies, to collect tolls and maintain the road.

In the meantime railroads were built, and local roads were made to reach their stations. Here and there sections of the National Road were abandoned, its milestones and some of its bridge masonry were appropriated by the inhabitants for building or other purposes, until a score of years ago it was almost impossible to find more than half the original route.

Within the past few years measures have been taken to restore the Old National Road. Nearly all of that section located in Pennsylvania has been incorporated in the system of State Highways of that state, and special appropriations made for its improvement. In Ohio and Indiana Good Roads Associations have taken up the matter, and are raising funds for accomplishing this result. So it may be that within a short time the Old National Road, for which congressional appropriations reached a total of nearly $7,000,000, will be restored to its pristine splendor and usefulness; not, to be sure, in carrying stage-coaches and trains of freight wagons, but for the passage of that modern device for consuming distance—the automobile.

The subject of the National Road should not be left until one point is made clear. Reference is always made to the fact that the road was built by the Federal Government. That is literally true. The appropriations were made by Congress, and the work was done by government officials. But there is another side to the case which is not so generally known:

When Ohio was admitted to the Union as a State in 1802, the law which provided for its admission to the Union contained a clause providing that 5 per cent. of the money received from the sale of public lands within the state should be used for constructing highways. Of this three-fifths should be applied to roads within the state, and two-fifths to the construction of a highway from the eastern boundary of the state to tidewater on the Atlantic Coast.

When Indiana was admitted about 1816, and Illinois at a later date, similar provisions were made; and it has not transpired, so far as current history shows, that the government of the United States has ever returned to those states the balance of the money due them for roads under these laws. Possibly if a settlement were to be made today, each of these states would find that they have enough money coming from the Federal Government to build many miles of the highways required by modern traffic.

Since about 1840 until about twenty years ago, a period of more than half a century, American civilization busied itself with building railroads. Highways were neglected, while twin streaks of steel were built into every corner of our great country, affording a development which, in extent of territory covered and the number of people permanently located, and the area of cultivation opened, has made a new record in the history of the world. So that we have now in the United States more miles of railroad than all the rest of the world put together.

While we were building railroads and creating an empire the nations of Europe were building highways. Now that there is a cessation in railway extension, Americans are giving their attention to the building of roads, for the comfort and convenience of our people, and for the reduction of the cost of transportation of food-stuffs and other products from the farms where they are grown to the market in which they are consumed, or to the shipping-point whence they are conveyed to such market.

The modern movement for good roads began in 1891 in New Jersey, Massachusetts followed in 1894, Connecticut in 1897, New York in 1898. Other states have followed until there are now forty-four states in the Union either with regularly organized Highway Departments, or with some method arranged for the systematic improvement of the highways.

About a dozen years ago a new factor appeared which has had to be reckoned with in the improvement of the highways. The factor is the automobile. At the beginning of 1904 there were approximately 50,000 of them in the entire United States. At the beginning of 1917 there are nearly 2,800,000 of them on our highways.

This is a new traffic superposed on the former traffic, as the use of horse-drawn vehicles has declined but little if any. The mechanical vehicles form just so much added wear on the roads, additional agents of destruction.

The broken stone or macadam roads of our fathers and grandfathers, which for a century and a half had been considered the highest standard form of construction, go to pieces rapidly under the combined travel of horse-drawn and motor-driven vehicles. The iron shoes of the horses and the steel tires of wagons break up the stone surface; then the low-bodied, swift-moving automobile comes along, and the dust which formerly was packed into the road and made it solid, is sucked out and thrown into the air, whence it is wafted away over the adjacent fields, injuring the crops and causing damage and irritation. Then the horses and wagons break up more stone, and the automobiles distribute it, and presently the whole road is ruined, requiring a new surface.

During the last eight or ten years highway engineers have been inventing new styles of road surface, and motor engineers have been inventing new and heavier styles of vehicles. First one has a little advantage, then the other. Motor trucks have grown in weight and carrying capacity until they exceed those of the railroad freight cars of a few years ago. When roads are built firm and solid enough so that they will stand the stress—with heavy foundations and tenacious surface—some other and heavier vehicle is invented. In their turn have come, in some localities, the auto-omnibus, carrying as many as thirty people, making regular trips between country towns; and motor freight trains, each with a heavily loaded motor truck in front, hauling a string of loaded freight wagons behind.

From other countries come stories of "trackless

trolley lines," with the wires strung over the highway and the cars running on the roadway without rails. Where the development of the vehicle is to find its limitations, or what the road of the future may be, no man can tell. It requires the best intelligence of today to design and construct roads which will meet the requirements of the present. The future must take care of itself.

In this country there is being expended annually from $350,000,000 to $400,000,000 for the construction and repair of the roads. There are about 2,200,000 miles of roads in the country, and rural mail carriers go over approximately half of them every day, carrying knowledge of the world and its affairs to 80,000,000 people who lived in comparative isolation ten years ago.

It may be stated that the automobile and the rural mail carrier are the two great factors of recent years in the demonstration of the necessity for improved roads and more of them.

Good roads mean to a people everything that is good. They mean higher standards of morality; better standards of thought and action; clearer comprehension of the great master-thought which we call nature's God. And they mean that freer intercourse which, particularly in our own country, has inculcated ethics and eliminated provincialism throughout the nation.

CHAPTER II

ROAD LOCATION

As the primary purpose of a road is to carry the traffic, it naturally follows that its location should be such as to accommodate the greatest amount of traffic. In classifying roads from the standpoint of location, therefore, three distinct groups appear. They are:

(*a*) Agricultural.

(*b*) Residential.

(*c*) Scenic.

While the roads composing these groups overlap, and one group frequently changes or melts into another, the subject must be considered from the primary necessities of the group itself, and provisions made for harmonious blending when two or more groups come together.

In the greater part of the United States the roads are constructed on section lines, one mile apart in each direction. When a section of country becomes thickly settled and the land is cut up into small farms, roads are often built on half-section lines, making parallel roads one-half mile apart, and sometimes closer than that. These smaller subdivisions usually occur near cities or towns, where the district may be given up to suburban residences or to market-gardening and similar purposes. In the older states and in hilly and mountainous regions the roads generally follow the valleys of streams, rising gradually to the higher elevations through the winding courses formed by nature.

But the grouping of roads remains practically the same. The greater the distance from city or town, the larger becomes the average size of farms. The industries of the people are more entirely agricultural. At the same time, in certain localities, great scenic advantages may be found which may affect the classification of the road. Thus, an agricultural road may have a residential character for a limited distance from the city line, then gradually become purely agricultural, and then blend into the scenic before it becomes a part of another road leading to another town or city.

In the cases of agricultural roads and residential roads most of the travel originates along or in the region immediately tributary to the road. In the case of scenic roads the major portion of the travel is attracted to the scenic section through which the road passes. The character of the traffic itself must be considered in determining the location.

According to the figures of the United States Office of Public Roads and Rural Engineering, the average haul of farm products in the United States is nine and four-tenths (9.4) miles, and the average cost of hauling from the farm to the market approximately 23 cents per ton per mile. As this cost average includes all classes of roads, improved and unimproved, it must be considered that in sections more remote from the market and where the roads are not improved the cost of hauling to market is so great that only the most profitable crops can be grown and marketed without loss. To this fact must be ascribed the abandonment of many farms some twenty or thirty years ago in Massachusetts, New York, Pennsylvania, and other states. In more recent years, since the building of roads has become a fixed policy, these lands have all been reclaimed, and now possess a high value as farm lands.

In locating a road through and from an agricultural community attention should first be given to the cultivable area which the road is intended to accommodate not only at present, but in the future. In older communities the selection of a road for improvement as a main road is largely a question of choice among existing roads, each having its special claims for preferment. The facts most necessary to a proper consideration are:

(1) Accessibility to the greatest cultivable area.

(2) Expense of securing correction of alignment, widening right of way, eliminating sharp corners (including necessary purchases or exchanges of small parcels of land), and avoidance of railroad grade-crossings.

(3) Cost of construction, including grading, culverts,

bridges, both on the road to be improved and those leading to it.

(4) The maximum grade and the distance to the city or shipping point.

(5) The probability or possibility of the road, after improvement, becoming a section of some through route between important points.

The relation of the first cost of construction to the various advantages of certain locations is always a delicate subject. Each individual case must be worked out by itself because no two cases are exactly alike. Where the location is under the control of state highway officials some of the factors are simplified. Where it is in charge of county or township officials, who must treat and negotiate with officials of adjoining townships or counties for the construction of a continuous road, much close figuring is required. The local officials are usually in close touch with the taxpayers, and taxpayers do not always think alike when immediate cost is considered with future advantages.

There is no question, however, that for the agricultural road the shortening of distance and the lowering of grades justifies additional expense. Marketing products must be figured on the cost of hauling per ton per mile. Slight grades enable larger loads to be hauled, and with shorter distance there is established a direct and permanent economy. This economy may justify

Fig. 1.—Road winding among the hills.

under some circumstances the location of diagonal main thoroughfares through a large area of section line roads. If the distance from a farm to market be 12 miles, 6 miles north or south, and 6 miles east or west, the diagonal distance would be less than 9 miles, and the relative reduction in distance would be the same at all intermediate points. The cost of marketing products would therefore be reduced by more than 25 per cent. from the region involved.

Whether the cost of the land for the right of way and the cost of construction would justify the establishment of this permanent economy is a matter for each community to determine. The proposition would resolve itself into a form about as follows:

"Will a 25 per cent. saving in hauling charges justify the cost of the new right-of-way which will reduce to that extent the hauling distance?"

From this cost must be deducted the cost of the improvement of the extra length of the longer route; but much of this saving will be taken up by building a road in a new location, instead of along the line of an old road. As a general principle, it may be stated that the construction of a main diagonal road through any considerable farming territory where the roads run at right angles, in order to make a short cut to the city or shipping point, is good policy; for the saving in cost of hauling is positive and permanent. The cost of an improvement so located, therefore, should not be looked on as an expense, but as an investment, which will continue to pay dividends as long as fields grow crops.

The differences in public sentiment in various communities and the differences in the laws of the several states have much to do with establishing roads in new locations, and for correcting defects in old locations. In some communities a large number of property owners will donate the strip of land necessary; then again an owner will be found whose avariciousness is such that he demands an exorbitant price for his property. In some states the condemnation laws are such that a great public improvement can be held up for years on account of the refusal of a single owner to part with the necessary strip of land on terms that his neighbors are willing to consider. In most states, however, the land may be taken under condemnation proceedings, the road built and put to use without reference to the length of time the question of compensation is held in the courts.

In a majority of cases the location and improvement of a main road through a community so enhances the values of the adjacent lands that the owner can well afford to donate the right of way. Exceptions occur in the case of small holdings, and where improvements such as buildings or orchards or other permanent improvements must be sacrificed. The local conditions are so varied that no specific method may be prescribed. The question is one of fairness to all interests involved.

In residential districts the problem of location is more simple. In many cases the district is a suburb of a city, and the location of the roads has been fixed by the original promoters of the residential colony. Where this was done several years ago, the central idea was probably accessibility to the railroad station, or to the interurban trolley line, which would provide transportation to the city. Since the popularization of the automobile this viewpoint has largely been changed, many suburbanites having their autos in which they ride regularly to and from their business in the city.

In this way the location of the main road leading from the suburb to the city becomes important. This importance is added to by the fact that auto-trucks, loaded with fuel and other heavy commodities, and auto delivery wagons from the city often take the place of the former railroad and local delivery system. Not only must the main thoroughfare be direct and of easy grade, but the character of many of the streets and side roads is changed. This raises the question of the location of roads which must carry the heaviest travel, and which for this reason must have the most substantial improvement.

The location of the residential road must always

control the character of the improvement. Where the most travel goes should be placed the strongest foundations and the most enduring surface. It is as wasteful to put a cheap and inferior pavement where a first-class one is demanded by the traffic, as it is to put a high-priced pavement on a back or side street where there is little travel. Nearly all of the more expensive pavements of today will go to pieces more rapidly under a very light traffic than they will under a heavy traffic.

An instance of this may be noted: The main street of a suburban village near one of the medium-sized cities was being paved. The street formed a part of the main country road leading from a large agricultural district to the city. The suburban village authorities decided on a high-class expensive pavement, both from sentiments of civic pride and economy.

A gentleman of large means had a palatial residence about half a mile back from the road just beyond the village limits. His private road branched off the main street at about the point where the pavement changed to the well-improved but less expensive country road.

The man, observing the paving of the street while the work was in progress, became impressed with its excellence, and decided that he would have his half-mile of road improved in the same way. He figured that if the street in the suburban village would last twenty or twenty-five years, under the traffic that it would be called on to carry, that his road, with its light travel, should last almost to the end of time.

The road was built according to specifications. That was twelve years ago. It was built with a concrete base and curbs, brick gutters, and sheet asphalt between the gutters. Today the street in the village is in first-class condition, carrying a heavy farm and local traffic, and with an exceedingly light maintenance cost. But the private road has only left of the original construction the concrete base and curbs. Five or six years ago the asphalt and brick were removed and a surface given of gravel mixed with some preparation which holds it in place, and prevents the formation of excessive dust.

The original surface went to pieces because of lack of enough travel to keep it in proper condition. The asphalt surface cracked and buckled from the cold and heat because there was not travel enough over it to keep the surface ironed down. Seeds of grass and weeds settled into the filler between the brick, and the growth pushed them out of place, let the water under them, and the frost did the rest. It was an example of a good pavement in the wrong place.

In locating a scenic road the basis for calculation is entirely different. The chief function of a scenic road is to attract travel. It may serve the further purpose of forming an avenue of transportation for an agricul-

Fig. 2.—Road cut into base of mountain.

tural section, but in most of the localities where roads are built for scenic purposes agricultural production is confined to occasional valleys or patches of arable land. Sometimes other purposes are served; such as getting out timber, moving mine supplies and products, etc.; but these various purposes tend rather to detract from the importance of the road than to add to it. The various industries could probably be better cared for by roads so located as to give them more direct routes and easier grades, with the scenic element left out.

The varied character of scenic roads makes impossible any set of hard-and-fast rules to be observed in their location, or grades, or types of improvement. "The Crest of the Blue Ridge" road, extending across the Carolinas about the top of that particular range of the Appalachian Mountains, requires one kind of treatment, and the "Columbia River Road" in Oregon requires another, yet both have the same purpose, that of attracting visitors and travelers from other sections of country to take advantage of the grand and impressive scenery which nature has established, of a favorable climate, and of such facilities in the way of places of entertainment as may have been provided.

There is a difference of opinion as to the maximum grade which should be permitted on a scenic road. Some authorities hold that it should never exceed 5 per cent. (5 feet elevation to 100 feet of distance), because that is said to be the steepest grade at which the average automobile can operate on high gear. Advocates of this theory affirm that the pleasure seeker prefers to travel a greater distance rather than shift to a lower gear while climbing a steeper grade. Others state that the grade, within reason, makes little difference, and that an additional charm is given to motoring when there are occasional hills to ascend or descend.

The fact that a location once made is likely to be permanent, and that in the course of years the scenic road is liable to cause the development of industrial resources of the region which will make the road available for other purposes, suggests that the grade be kept down to the 5 per cent. limit wherever possible without too great a sacrifice of view, or extension of distance, or the overcoming of other special disadvantages. It frequently occurs that in maintaining a reasonable grade the location must be made along the sides of mountains, sometimes by a zig-zag course; streams and chasms have to be crossed by bridge or trestle, and many other unusual conditions encountered. These must be studied individually, as no two instances are nearly enough alike to come under any general rule that may be laid down.

A qualifying factor in all road location is the question of through routes. In some states this is covered by state laws, establishing State Highways connecting important centers. In other states and sections of the country Road Associations of greater or less importance have a definite influence on local road locations. Such organizations as the Quebec and Miami Highway Association; the Lincoln Highway Association; the Dixie Highway Association; the National Old Trails Association, and many others of either national or sectional importance, have been assiduous in their efforts to enthuse local authorities to a point where they will locate and improve the particular sections within their respective jurisdictions. Much force and impetus have been given to the Good Roads Movement by the work of these associations, and in the general discussion of location of roads the differences between roads of a purely local character and those which will accommodate general travel have been clearly brought out.

It may be assumed that a through route traversing a county or other subdivision should be located where it will accommodate the greatest number of the local population and the greatest amount of local production. Any other basis of calculation would give the road more importance as a through route than as a local necessity. With the exception of a very few roads between very large cities, and a few purely scenic highways this is not a fact. While the local usefulness of the highway should be the first consideration, careful attention to the advantages of a constantly growing

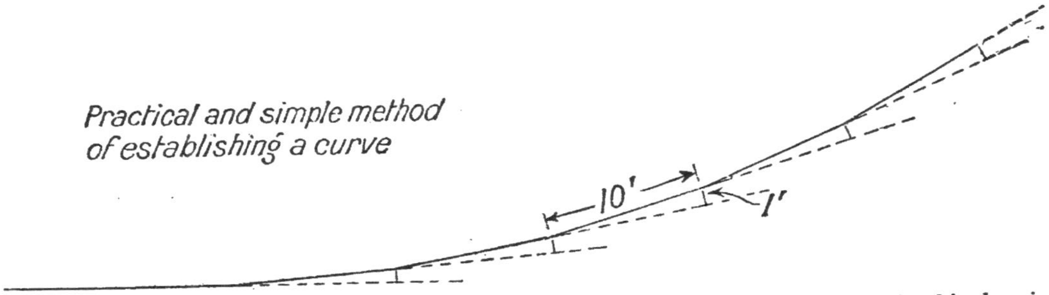

Fig. 3.—A simple and practical method of establishing a true curve is shown in this drawing. From the point where the curve begins continue a line straight ahead 10 or more feet as shown in the dotted lines. Then measure 1 foot, more or less, according to the curve desired, and draw the center line from the original starting point through and beyond the new point, continuing the operation every 10 feet until the end of the curve is reached.

through travel, and to the proper connections at county or other divisional lines, will contribute to the highest usefulness of the road.

In locating any new road, or in relocating a road for the purpose of securing better grades, or for other reasons, modern traffic demands that provision be made for a "line of sight." By this is meant the distance at which vehicles approaching each other may be seen by the drivers of each.

A trifling computation shows that two automobiles, each running at 20 miles an hour, approach each other at the rate of $\frac{2}{3}$ mile or 3520 feet a minute. Even if drivers were always alert, at least half a minute, or 1760 feet, is the smallest possible amount of time and space in which safety can be assured. It becomes necessary, therefore, that in locating a road a view of at least $\frac{1}{3}$ mile ahead be provided for, either by sight along the road or across the intervening fields. Where this is not possible by reason of hillsides or forests, or buildings, or other obstructions, cautionary signs of an unmistakable character should notify drivers to reduce speed.

In general road locating this provision is not as difficult as it may appear at the first glance at the subject. Every sober driver knows enough to slow down at a sharp turn and keep to the right side of the road; but with 2000 feet of road in sight, more or less, the driver is likely to take chances, or to be extra cautious, according to disposition. Every one who has driven a motor car will appreciate the distinction.

To so locate a road that the line of sight will be extended to its greatest possible limits is of grave importance whether the travel be horse drawn or motor driven, or both. In but few states are horse-drawn vehicles obliged to carry front and rear lights, which fact adds an extra hazard at night, even under the most favorable conditions. Common prudence, however, requires that a driver be able to see an approaching vehicle at a distance which will enable him to take reasonable precautions for safety. To accomplish this the highway officials under whose jurisdiction the road is located must constantly keep in mind the importance of the line of sight.

The great importance of the correct location of a road lies in the fact that it is the one element of the road that is expected to last for all time. Grades may be destroyed; surfaces will wear out and require replacement; bridges and culverts are subject to damage; but the location always remains, either as a monument to the intelligence and integrity of those who placed it properly, or as a constant and permanent reminder of the inefficiency of those who may have been responsible for an improper location.

This is especially true where a road is built with the

proceeds of a bond issue, and particularly in cases of long term bonds. The location and the right of way, taken together, are the one tangible basic asset which does not require maintenance or other expenditure. If the location has been wisely made, its value to the community increases year by year, in the ratio that the prosperity of the people is developed; and when the bonds are paid off and cancelled, the people are rich in the possession of a road location on which can be placed or kept whatever surfacing the needs of a constantly changing traffic may require.

A certain road in the interior of the state of New York was located on a line where two land grants joined. At one point the road passed over a high hill, the distance from the bottom of the hill on one side to the bottom on the other being about $\frac{1}{3}$ mile. The apex of the hill was nearly 200 feet above the level of the roads on either side, forming grades in some places of 20 per cent. and upward.

Over this hill was hauled the marketed products of a considerable farming area, perhaps thirty or forty good farms, for more than a century. Three or more generations of farmers hauled small loads or "doubled up" to get their produce over the hill. It was something more than one hundred years from the time of the original location of that road before the community became sufficiently interested to demand a change.

It required two or three years of almost constant arguing by some of the younger men of the neighborhood before action could be taken. Then a road was built around the base of the hill, practically level, and the old hill road abandoned.

It fell to the lot of one who had been born and reared in that neighborhood, but who was making his first visit there in more than twenty years, to take a tape line and measure the surface lengths of the two roads. He found that the new level road around the base of the hill was exactly 49 feet longer than the one over the hill.

He pondered. He recalled the struggles of his youth; the struggles of his father, and grandfather, and great grandfather, and their neighbors in the century work of advancement toward prosperity and education and higher civilization. He computed roughly the annual farm production which had gone over that hill; the time wasted; the small loads enforced; the broken hame straps and whiffletrees, and a thousand other factors brought up by memory.

Then he considered again that 49 feet in a total distance of about 1800 feet of hill and level road. Then he wrote for a book a chapter on the subject of "Road Location."

CHAPTER III

ROAD GRADES

For purposes of discussion the subject of road grades must be separated into two phases. One phase has reference to a departure of the surface from a level; the other deals with the road-bed itself.

A road is up-grade or down-grade whenever it is not level, and the measurement may be made from either end of the incline or decline. Beginning at a low point and continuing the survey to higher ground, the up-grade may be computed in a percentage, which means a given rise in each 100 feet of distance. It may be stated that while there is no especial reason except, perhaps, uniformity of expression, it is an almost universal custom to figure grades on the ascending instead of the descending scale of measurement. Thus a grade of 2 per cent. means a rise of 2 feet in each 100 feet of distance. Grades are "light" or "heavy" according to the amount of rise per 100 feet figured as a per cent.

It is a general practice among road builders of experience to avoid an absolute level in laying out a road for any considerable distance. One of the reasons for this is that as there must be a slope to the side ditches to enable the water to run off, the ditches would be too shallow in the center of the level strip and too deep at the ends where the water is to be discharged. It is considered better practice, in establishing the grade to make the slight inclinations in the road surface so as to keep the ditches at a uniform depth. Besides, in a region so flat that a nearly level road becomes necessary even a slight grade assists in securing a prompt run off of surface or storm water.

In most states where state highways departments control, road grades are limited to a maximum of 5 feet to the 100, or 5 per cent. Exceptions may be made in rare instances for short distances, or a fractional increase may be permitted where the lay of the land is such that a large saving in cost or distance may result. These cases are rare, however, when compared with the total number of miles of roads improved. Many city streets have steeper grades, due to the fact that cities were laid out on irregular and hilly surfaces, and improvements made before the necessity for lower grades became apparent.

The principle involved in fixing the limit of steepness (the maximum grade which may be allowed on a road) is that over which the greatest loads which can reasonably be expected may be hauled at the least expense. In studying this problem it is better to use horsepower because most motor-driven vehicles are so constructed that with change of gear they can negotiate almost any ordinary grade.

The weight of load that a horse can pull on a level road or on any grade depends on the surface of the particular road. So for general purposes it is better to consider the pulling power, or tractive force, of the horse as compared with the grade.

Actual tests have shown that a horse working steadily can exert a steady pull of 120 pounds. For a short time, as in starting a loaded vehicle or pulling up a short hill, this may be increased up to a possible 500 pounds. Suppose the road surface is one on which a pulling power of 120 pounds will haul 1200 pounds of load on a level, it has been found that only half this load, or 600 pounds, can be hauled up a 6 per cent. grade, and only one-fourth, or 300 pounds, up a 10 per cent. grade, with the same horsepower. The exception of the short distance, or the short space of time in which the horse can exert greater pulling power, should only be given application when it is absolutely necessary to do so.

A factor which enters largely into the determination of the maximum grade is the direction in which the heavy loads travel. If it be an agricultural road, and the city or shipping point is in the valley with the farms accommodated by the road located on the adjacent hills, it is reasonable to suppose that the greater part of the heavy traffic over the road will be down grade. In such cases a steeper grade is permissible than in localities where the railroad and shipping points are high up on the side of the hill, with the farms in the valley. In the latter case the heavy loads must be hauled up hill, and grades should be made as easy as the topography and the elevation will permit. Care must be taken when the heavy traffic is down grade not to have the grade steep enough so that brakes on wheels will cause injury to the road surface.

Authorities differ greatly as to the steepness of the grade on which various surfaces may be used. An earth road has been accustomed to be considered as available at any grade which was possible of ascent or descent. In certain hilly regions very steep hills are traversed. Grades up to 20 or even 25 feet to the 100 are employed and kept in fair condition. When a country road becomes too steep for any hard surfacing, the only alternative within the limits of reasonable cost is to leave it with a surface of earth or gravel which will not wash badly. On steep city streets a pavement of stone blocks or cobble-stones is usually used. A cobblestone pavement is reported on a street in San Francisco which has a grade of 51 per cent.—51 feet rise in 100 feet of distance.

Macadam and gravel roads have been used on almost any grade. There are numerous examples of their employment on grades up to 18 or 20 per cent. Asphaltic macadam may be used up to a grade of 7 or 8 per cent.; asphaltic concrete, up to a grade of 5 or 6 per cent.; vitrified brick, up to a grade of 6 or 7 per cent.; concrete, up to a grade of 4 or 5 per cent. Special forms of brick and special corrugations in concrete, both made for hillside use, may add to the steepness of the grade where they may be permitted.

These figures are relative rather than arbitrary. They may be modified by any one of a variety of local conditions, and especially by the character of the traffic. Wide and narrow steel tires, pneumatic and solid rubber tires, the weight of the heavy loads, and other factors must be considered when determining the maximum grade of a road on which a certain surfacing is to be placed; or determining the proper surface for a road of a given grade.

Further modification of the figures may depend on the details of construction. If the surface be made somewhat rough, or so that it will wear rough, by placing reasonably large sized stones in the top course a better foothold will be given than when the surface is of fine stone or fine gravel. With the better foothold a slightly steeper grade is permissible. In the determination of grades, however, the primary thought must always be to keep the percentage as low as possible.

In establishing road grades consideration must be

given to the level of intersecting roads; to the level at which bridges and culverts must be constructed, and to the general scope of the road itself. When a road is being built for a distance involving several miles with an ascending grade, good practice does not permit, except under extraordinary circumstances, a down

Fig. 4.—A long side-hill grade in Ohio.

grade at any point in the road. An instance of this was presented at a road convention not long ago, when it was stated that in building a road from sea level to the crest of the Blue Ridge Mountains, a rise of some 5000 feet, no steeper grade than 4 feet 6 inches to the 100 feet had been employed, and no down grade whatever permitted along the ascending route. Of course this does not apply to bridge approaches nor to conditions made necessary by railroad crossings, road crossings, or other purely local factors. As the general trend of the road was toward a higher elevation, the grade of its different sections must conform to the general plan.

In establishing road grades special treatment of curves is necessary. When a curve is approached there should be a gradual reduction of the grade so that when the curve is reached the grade should not be more than half as steep as on the straight road. For instance, if the grade of the straight road is 4 feet to the 100 feet, on the curve it should not be more than 2 feet to the 100 feet of distance. This rule applies to any ordinary curve. If the curve be a short one the grade should be reduced still more, and on very short curves should be reduced to a level.

By a "short curve" in this connection is meant the degree of curvature, or the radius of the curve; not the length of the road on the curve. A curve is a part of a circle, like a part of the rim of a wheel. The radius of any curve is the distance from the rim to the center of the circle, as to the center of the axle on which the wheel revolves. Thus a radius of 100 feet would mean that the curve would be a short section of a circle 200 feet in diameter. The measurement is usually made to the center line of the road, and the road should always be widened on a curve. The best authorities seem to consider that the widening should be on the inside of the curve, keeping the outside to a true curved line parallel with the center line of the road.

On the national roads of France the radii of curves are established at a practical minimum of 165 feet, while on communal or country roads curves as short as 50 feet radii are permitted. In the measurement of curves all engineers do not seem to follow the same practice, some measuring from the inside line of the road, others from the center line, and still others from the outside line. This can be best illustrated by the turn of a street where the curb comes to a corner. If the inside measurement were taken, the radius would be zero, as the same point would be the beginning of one straight section and the end of another. If the

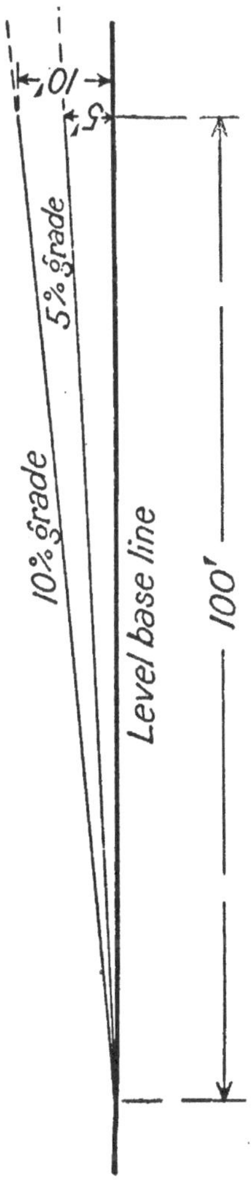

Fig. 5.—Method of determining the grade of a road.

measurement were made on the center line, and the street were 40 feet wide, the radius would be 20 feet; and if the outside line were taken as a basis for measurement, the radius would be 40 feet, the full width of the roadway. On country roads, however, sharp turns rarely become necessary, the usual right of way being wide enough to make a reasonable curve.

Figures by engineers show that on a 12-foot road the radius of a curve must be at least 100 feet to accommodate a four-horse team and long vehicle with a total length of 50 feet. If the road is 18 feet wide the radius of the curve can be reduced to about 66 feet and still keep the rear wheels on the roadway. On most roads, however, it is not the extra long teams and vehicles which must be provided for, but the facility and safety with which shorter ones may pass each other.

In the construction of the road-bed to which the term "grade" is usually applied, several factors must be considered. If the grade is to be built up, as an embankment on low ground, or if it be necessary to make a cut through a hill, the character of the soil and the amount of earth to be moved are the important features. Where a hill and a hollow come together the matter is simplified, as the earth taken from the hill can be dumped into the hollow with the least possible expense. When there is a haul of any considerable distance, either to dispose of earth taken from a cut or to bring earth for a fill, the expense grows rapidly in proportion.

By studying the cost accounts of a large number of contractors it has been found that in moving earth from a cut to a fill drag scrapers can be used to the best advantage when the distance does not exceed 100 yards; after that wheeled scrapers are best up to a distance of 300 to 400 yards. For longer distances dump wagons are more economical.

In making the fill the earth should be deposited in layers of approximately 8 to 12 inches in depth. Each layer should begin next to the hillside and be built outward as far as the earth from the adjacent cut is intended to be used. The first deposits of earth should be along the center-line stakes of the road, and a slight slope from the center to the sides should be maintained. The bottom or base of the fill must be made to its full width to secure the best results. The some-time practice of building up the center and dumping earth over the sides of the fill to complete it to the proper width has not been found satisfactory.

Where it is intended to complete the road during the same season a heavy road roller is necessary to pack the earth as each layer is placed. An important fact to be considered in this connection is, that while loose earth at first occupies a greater space than it did in its original position, when compacted, either under heavy rolling or by settling when time be given, it will occupy much less space than it did in its former bed. For instance, 100 cubic yards of measured earth taken from a cut, placed in an embankment, and so compacted as to make a proper fill for a road, will shrink to about the following figures:

One hundred cubic yards of gravel or sand will compact to about 92 cubic yards; of clay, to about 90 cubic yards; of loam, to about 88 cubic yards; and of loose surface soil, to about 85 cubic yards. Clay if subjected to the action of water will have a much larger shrinkage. These figures should always be taken into consideration in providing the earth necessary to make a given fill or embankment. They apply alike whether the earth is compacted by heavy rolling, or whether it is left over for a season to settle before completing the road, which is a custom in some sections of country.

The side slopes on fills and cuts must be at such an angle as to insure stability, and at the same time not waste material or labor. When the same earth is used, the slope on the sides of the fill may be figured the same as the side slopes of the cut from which the earth was taken. The tendency of the earth to wash into small gullies or to wash or settle away under storm or frost conditions must be considered.

In general, the slope is figured by the foot on the level compared with a foot of elevation. The usual practice is $1\frac{1}{2}$ to 1, which means $1\frac{1}{2}$ feet on the level to every foot in height, in most classes of earth. This is usually employed with sand and gravel. Some kinds of earth will stand a slope of 1 to 1 when it is firmly packed. Some kinds of clay will not stand unless the slope be 4 or 5 feet on the level to 1 foot in height.

The earth should be spread over the entire width required at the beginning of the work of making the fill. Take a 24-foot roadway as an instance, to be placed on a fill 10 feet high. If the slope is to be $1\frac{1}{2}$ to 1 the bottom course of earth should be 54 feet wide, allowing the 24 feet for the roadway and 15 feet for the slope on each side. By depositing the earth in this manner, keeping the fill a trifle higher at the center than at the sides, rolling close to the edges, and making each successive layer of earth narrower, the best results are obtained.

In every instance the factor of common sense (horse sense) on the part of the contractor or official in charge of the work must be reckoned with. An instance in point may be mentioned:

It had been decided to widen an existing road from 24 to 40 feet. In one place the road was built on a fill for a distance of about 300 feet, the fill being about 30 feet high over the lowest part of the gully, and tapering gradually to the road level at each end. In widening the fill and to provide special stability a new fill, 20 feet in width, was built along and against one side of the old one. The culvert was lengthened, the earth placed in position, and compacted with a heavy roller.

The old embankment had been in place for several years, and the slope had accumulated a heavy sod, and had become covered with a dense growth of rank weeds. Common sense should have taught the official in charge the necessity of taking a side hill plough and cutting off that sod, and removing or burning the weeds, so that the fresh earth should be made to amalgamate as closely as possible with the compacted earth of the old fill. This was not done, however. The fresh earth was deposited on top of the sod and weeds, rolled as solid as possible, and the road surface put on.

The next summer a lengthwise crack appeared in the surface of the road about where the junction of the new and old fills was located, and that portion over the new fill settled slightly. The officials were still discussing what steps to take to protect the road when the fall rains came, and it was decided to let the matter go over until another season in the hope that the difficulty would adjust itself. The second summer there was an opening in the surface averaging a foot in width and the surface over the new fill had settled 4 to 6 inches, requiring practical reconstruction.

Investigation showed that the sod and other vegetation in the course of decay had formed a thin partition line between the old fill and the new; and that partition was of such a character that the new or outer fill had been caused to slide downward and outward to an extent which had practically ruined the surface. The difficulty was overcome at considerable expense and no one connected with that job is likely to commit the same error again.

In building a road grade across a low marshy section of ground it is wise to make a careful investigation of the subsoil and of the earth strata underneath. It sometimes occurs that underneath the marsh is a thin stratum of clay perhaps only a few inches thick in places, and underneath the clay a layer of either very soft mud or quicksand. If the fill is to be a heavy one, so as to bring the road grade well above the adjacent marsh, it is very likely to break through the layer of clay, and form what for want of a better name is termed a "sink hole." Road builders in many sections of the country have encountered these conditions, and have spent large amounts of money putting on additional earth, which would soon settle again, making more earth necessary.

While each individual case must be treated according to the conditions found to exist, a number of contractors have reported that in their experience it is safest and most economical to find out the depth of the clay or hard earth or even soft rock underlying the low ground by cutting holes in it at regular intervals along one side of the road location; then determine the character of the earth beneath with an iron rod. If it be of a character that would develop into a "sink hole," it has been found cheapest in the long run to drive a row of piles close together on each side of the base line of the proposed grade embankment. The particular spots where this is found necessary are not usually very large in extent; and if the piles are driven down to the hard earth underneath a complete protection of the fill is established, and the settlement is likely to be very slight

In locations where the grade is built in a cut there is a variety of conditions. Sand or gravel; earth, with or without being mixed with stone; hard-pan; rock, soft or hard; boulders, from sizes which can be easily handled to those which must be blasted to pieces; clay, ranging from that which will dissolve easily in water to that on which water has but little effect—these are all subject to different treatment. In some cases the plough and scraper, as previously stated, will be found economical. If the cut be fairly deep, a steam shovel may be used; and on jobs where there is a large quantity of earth to be moved the portable railway and dump cars are employed. Where rock or heavy boulders are encountered a derrick of some sort is a money saving part of the equipment. Even a portable hand derrick, operated by a windlass and cranks, will often effect a saving in the cost of loading rock for removal. Generally the equipment should be consistent with the work, so that no part of it should be idle and creating expense while other parts are in use.

It has occurred frequently that there were more ploughs to loosen the earth than scrapers or wagons to take it away; or that there were more moving facilities than there was earth ready to move. Steam shovels with expert and high-priced operators are often idle part of the time because of a scarcity of wagons or dump cars. On the other hand, it sometimes occurs that there are too many wagons. Where drill work is going on for the blasting of rock, the drills can proceed without loss of time by going ahead into new work, but the arrangements for removal must be carefully proportioned to avoid useless expense.

What is known as a "vertical curve" is employed in changing from one grade to another. If a section of road has a grade of 2 feet to the 100 feet a point may be reached, as at the foot of a hill, where the grade will be increased to 5 feet to the 100 feet. In making the change, so that it will not be abrupt from one grade to another, it should be made gradually; and at the top of a hill the up-grade should be reduced gradually to a level, and the down-grade on the other side approached gradually. Where property damages are involved, it may be well to have these "vertical curves" scientifically determined. For all ordinary purposes good common sense may be depended on to make a change from one grade to another without abruptness.

In building up a grade by a fill or embankment, any rock or large stones should be used at the bottom, with earth above. No large stones should lie near the surface. The reason for this is that earth never packs so hard but that it will settle to some extent; and large stones, forming a rigid substance, will, if near the surface, make the surface irregular, and distort the surfacing material in the spots where they occur.

The principles involved in establishing and constructing a road grade relate principally to keeping the road from being level or flat, and from having a grade of more than 5 per cent.; constructing the grade so that the proper gradient will be maintained in cuts or on fills, or on a level where a level is necessary, and combining the level with the grade without abruptness. Then the building of the grade, either by embankment or cut, so as to maintain the roadway in proper condition to carry the surface that is to be placed upon it.

Given the location of the road, the center line of the road is easily established. This should be perfectly straight as far as the road extends in one direction. Whether the gradient be ascending or descending, the center line stakes should continue in a straight line until a curve is reached. The side ditches, the fills, the cuts must be measured at the proper distance from the center line. The slopes of the fills and the outer slopes in the cuts must all be measured from the center line of the road.

In making contracts for building road grades it sometimes occurs that one contract ends at about the edge of a hill through which a cut is to be made. The adjoining contract may be over low ground where a fill is necessary. In such case each contractor would naturally make his bid on the amount of earth to be removed or to be supplied. Unless it be by pre-arrangement, where the two contractors may work together with a proper reduction of cost, the community will find itself paying two prices: one for the removal of the earth from the excavation or cut, and the other for the cost of earth to be put into the low ground to make the fill. By having the cut and fill under one contract, and as nearly as possible covering the same amount of earth excavated from one point and filled in at another, with a single cost for the movement of the earth, much money may be saved.

CHAPTER IV

ROAD DRAINAGE

Drainage is considered by most road builders the most important factor in the construction of any road. Water may be considered the most destructive agent which can be encountered in building any road and in keeping any road in condition. Provisions for taking the water off the road surface and from underneath the road, and getting it entirely away from the roadway, must be comprehensive and complete if the road is to remain in good condition.

There are two distinct phases of the road drainage problem: One is the taking care of the surface water, storm water, melting snow, etc., getting it quickly into the side ditches and carried away through proper drainage channels. The other relates to the subdrainage; getting the water out of damp or wet soils or subsoils, so that the foundations of the road may be protected. In the final discharge the waters from above and below are usually carried off through the same channels, though it sometimes occurs that each may have an independent outlet.

Surface drainage begins with the slope or crown of the road surface. The height of the crown and its slope differs with each type of surface, and are specially mentioned in the various chapters devoted to road surfaces. In practically all country roads which are given hard surfaces there are shoulders or "berms," as they are sometimes called, extending from the hard surfaced portion to the side ditch. As a general rule these shoulders maintain the same slope as the hard surface, unless there be some reason for changing it.

In building any hard surface road the shoulder should be provided for either by excavating into the solid earth for the roadway, leaving the shoulder standing, or by building up a shoulder with good, firm earth, and rolling it to a solid condition. When the road is built care must be taken that the road material and the shoulder are packed firmly together; otherwise water will seep into the joint and cause damage either to the roadway or to the shoulder, or to both.

The shoulder should never be higher, not even by the fraction of an inch, than the hard surface of the road. If sod is permitted to form on the shoulder, care must be taken that it is kept down so that it will not impede the free flow of water from the road surface to the side ditch. In some soils, where the water has a tendency to soak into the earth, it is well to encourage the formation of sod. In a specially sandy region, where the sand is of a quality that when dry it is likely to be carried away by the wind, special kinds of sod-forming grasses are encouraged for the protection of the shoulder itself. But if the soil is such that the water will run off readily, it is better that the sod be not permitted to form or other vegetation to grow.

Over the shoulders the water gets to the side ditches of the road. Or, if the road be on an embankment or fill, the water spreads to the adjacent fields and finds its way into the natural channels of the drainage of the region. It can be readily seen that if the level of the road is materially higher than the surrounding country there is no need for side ditches except perhaps at the bottom of the embankment, to direct the water to the proper outlet. If the road is in a cut it must have a side ditch on each side, the same as if it closely approximates the general ground level. If it is on a side hill it must have a side ditch on the side next the hill, the water from the outer side draining away naturally.

In the past there have been three kinds of ditches which seemed to have popular approval: the wide flat ditch, the V-shaped ditch, and the ditch with nearly vertical sides and flat bottom. Each has been used whenever, in the judgment of the officials, the nature of the soil would permit their use. Modern practice, however, favors the broad flat ditch for a number of reasons. It will carry more water without developing a swift current; it is not subject to so much wearing away by the current nor to obstruction by cave-ins of the earth at the sides; and whatever moisture is left dries up more quickly because of the greater area of exposure to the air and sun. Besides these reasons which apply to the drainage duty of the ditches, there is the safety of traffic on the road to be considered. If the ditch be wide and flat the road shoulder is usually graded gradually down to the bottom of the ditch, so that a vehicle running into the ditch is not in danger of capsizing or other damage; while with deep narrow ditches a vehicle going at any considerable rate of speed which inadvertently gets a wheel into the ditch, either in darkness or by defective steering or apparatus, or through plain carelessness, meets with disaster. Not infrequently have lives been lost by reason of narrow deep ditches, which might otherwise have beensaved.

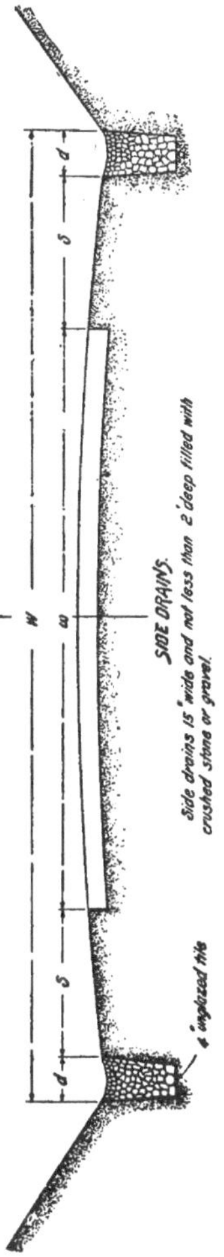

Fig. 6.—Loose stone under-drains below the side ditches.

When the road is built on a side hill, or where there is any considerable area from outside the right of way which drains toward the road, back-ditching often becomes necessary. This consists of cutting a ditch parallel with the road a short distance up the hill, which will receive the surface water from the area above and

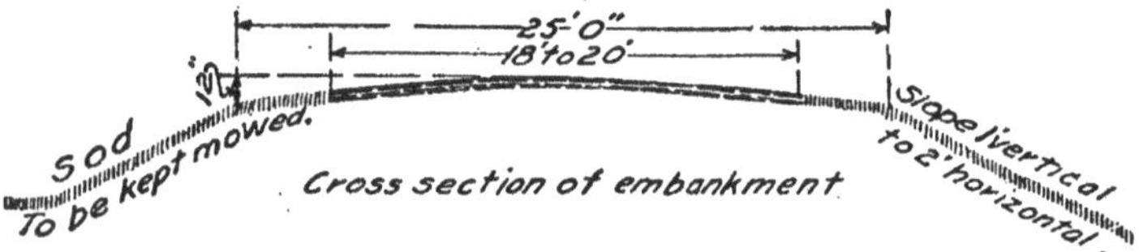

Fig. 7.—Roadway drains itself.

carry it to the point of delivery. This back-ditch keeps the heavy volume of storm water from flooding the side ditch of the road, and also prevents the wash of sediment into the road ditch.

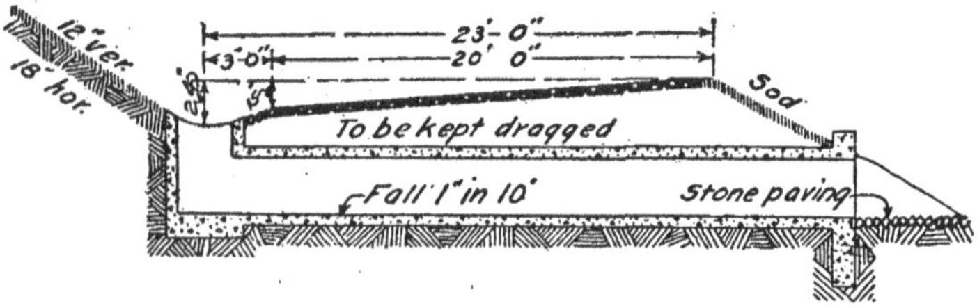

Fig. 8.—Culvert under a side-hill road to carry the water from the ditch on the upper side.

There is much difference of opinion as to the minimum and maximum grades at which a side ditch may be constructed. Some road builders assert that a grade of 2 inches per 100 feet is the least grade that should be employed. Others maintain that a properly constructed ditch having a fall of ½ inch to the 100 feet will carry off the water under all ordinary circumstances. So slight a fall, however, is not likely to create a current that will permit the ditch to keep clear of sediment. This is quite an important feature in localities where litter, in the form of dead leaves or other vegetation, is likely to be blown by the wind into the ditch. Unless the grade of the ditch is steep enough so that the

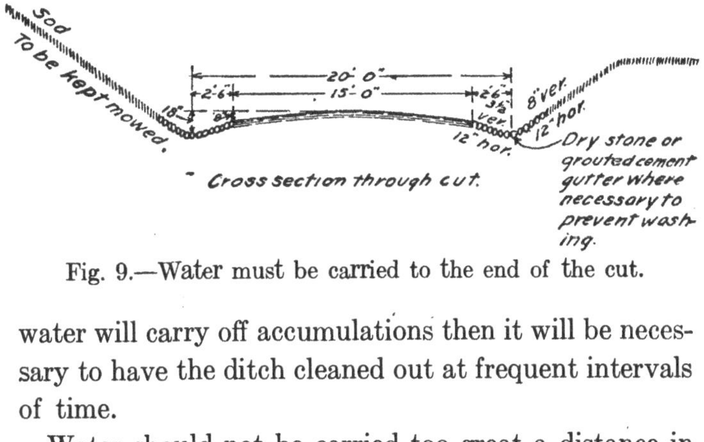

Fig. 9.—Water must be carried to the end of the cut.

water will carry off accumulations then it will be necessary to have the ditch cleaned out at frequent intervals of time.

Water should not be carried too great a distance in side ditches before being given an outlet which will take it off the right of way. The distance that it may be carried, however, depends on so many conditions that much must be left to the judgment and common sense of the road builder. If the grade be nearly level, not more than 2 to 4 inches to the 100 feet, the water can be carried several hundred feet without damage if provisions are made for taking care of the volume of water which may come from heavy rains. Such provisions, when the water cannot be taken away within a shorter distance, usually consist of making the ditch wider and deeper toward its lower end.

Take a sudden rainfall where an inch of water may fall within a couple of hours, with a road 30 feet wide between the centers of ditches, 15 feet from the center of each ditch, there will be 15 inch-feet, or $1\frac{1}{4}$ cubic feet of water to be carried off for every foot in length of road. One hundred feet of road would shed into each side ditch 125 cubic feet of water. At 200 feet this would be increased to 250 cubic feet, and so on in proportion, until an ordinary side ditch would be taxed to or beyond its capacity.

If the road and ditch be on a heavy grade the proposition requires no less study and judgment. With a swift current and a constantly increasing volume the water is likely to wash out either the bottom of the ditch into a deep gulley or wash out the side of the road, or both. This is particularly true if the earth is soft or sandy; most alluvial soils are easily washed, and in all such soils care must be taken that the water does not obtain either too great a volume or velocity. Where the earth at the bottom of the ditch is of hard-pan, or hard gravel, or some kinds of hard clay or hard loam, it will stand a good deal of rapidly flowing water without much damage. It must be the duty of the road builder when constructing the ditches to thoroughly inform himself as to the character of the earth in which he is working, and make his plans accordingly.

Where there is a heavy grade there is almost invariably some means of getting the water away from the road at frequent intervals, at most every few hundred feet, and turning it into some natural water-course or low place where it can be turned into natural drainage channels. No water should ever be carried in a road side ditch further than to a point where it is possible to dispose of it off the right of way, except where damage to private property is involved. When such a drainage-point is reached the water must be taken through a culvert or pipe from the upper to the lower side of the road and turned into the drainage channel. The culverts must be of sufficient size to carry all possible volumes of water to which they may be subject, and placed at such a pitch that they will not only carry off the water, but will keep themselves washed clean of sediment.

When water is turned over the side of a bank from the outlet of a culvert or ditch, it may be necessary to protect the bank so that it will not wash out. This may be done by making a trough of planks, or by building a stone wall with the stones sloped outward and downward—on the same plan that shingles are placed—or, what is far better if the means permit, building a concrete or masonry channel from the end of the culvert or ditch to the waterway below. The fact must be kept in mind that even a small amount of water, running down a steep bank, without protection to the bank, will wash it away gradually; and when a flood of water comes the wash is likely to be great enough to undermine a considerable portion of the road if the bank is not properly protected.

When the water is carried through a culvert or pipe from one side of the road to the other the culvert or pipe should not be squarely across the road, but at an angle which will carry the water in a down-hill direction. The opening to the culvert from the inside road ditch, if on the same or an increased down grade, should be curved, so as to avoid a direct heading for the water. If, as frequently happens, the culvert may be placed at a lower level, so as to extend underneath the embankment, so as to discharge the water near the bottom of the bank on the outside, it may be necessary to construct what is known as a "drop inlet." This drop inlet may be made in a variety of ways and from various materials, depending somewhat on the depth. Concrete or stone laid in cement mortar are usually considered the most durable, and therefore the most economical in the long run. The drop inlet should extend a few inches below the bottom of the culvert opening for the double purpose of catching matter which might clog the culvert, and for furnishing a "water cushion" for the falling water. The size of the drop inlet must be gauged like the ditch and the culvert so that it will carry any reasonable amount of water which may reach it. It should be protected at the top by a grating.

Occasionally instances are found where for several hundred feet on a fairly steep grade there is no practical way of getting the water away from the road without heavy expense. In such instances some excellent road builders have laid sewer pipe underneath one of the side ditches, with openings into it every 200 or 300 feet. The water from the other side of the road is carried across in culverts at similar intervals and turned into the pipe which carries it to the bottom of the hill or to the waterway. In this way the side ditches are relieved of the volume of water before it becomes great enough to do much damage.

There are few sections of country which do not furnish a natural outlet for drainage; and in nearly every section there can be found records of rainfall and drainage areas of given streams. These data should be consulted in determining the size of ditches and culvert openings. Often the most accurate figures are those found in the offices of the engineering departments of railroads, and reliable topographical surveys have in many cases been made by state or federal authorities.

In every instance the purpose of the surface drainage of a road must be kept in mind. The slope of the road from the center to the side ditches; the grade of both the road and ditch; the entrance to the culvert, the culvert and the discharge therefrom; the protection of the bank; the use of sewer-pipe under the ditch; the connection with a natural waterway—all must be parts of the plan to take away the surface water from the road, with the least expense and with the greatest permanency of construction.

Subsurface drainage, while equally important to the life of a road, has in many cases been given less attention than it deserves. Some otherwise good road builders have overlooked the fact that it is as necessary to get water from under the road as it is to get it off the surface.

Climatic conditions and the nature of the soil are the most important factors in subdrainage. When a road is built on an embankment of sand or gravel, or of soil in which sand or gravel predominate, there is no need for subdrainage. Under these conditions the earth drains itself. The same is true if the road is on fairly high ground with a similar soil. But most soils other than these retain water in their composition unless there is a direct means of draining it off into regular drainage channels. Soils which consist largely of clay or other plastic material are liable, unless properly subdrained, to become practically impassable in wet weather, and nearly every soil except self-draining sand and gravel is likely to heave with the frost and become honeycombed when the thaw comes. It is to rectify these conditions that subdrainage becomes necessary.

In low places ponds of water along the roadside, if permitted, are likely to keep the subgrade saturated. Of course, this water should be carried off by the surface drainage provisions, but this does not always seem to be possible. In such cases subdrainage must be put in for the protection of the road. Sometimes springs from hillsides or from ground higher than the road will keep the earth under the roadway saturated for a considerable distance. Subdrains are necessary to take the water away. There may be a stratum of rock under the roadway through which water may seep or the rock may be so shaped as to hold water, which will be very damaging to the road if not removed.

Some kinds of loams hold water almost as readily as clay; and the clearest guide as to what soils subdrainage is needed in is to observe the amount of expansion when the ground freezes in winter. The expanding of the earth loosens the particles from each other; then when the thawing process takes place—and the ground thaws from the bottom as well as the top—the soil, softened by the water, is churned up by the travel, and it takes days and sometimes weeks for the road to dry out. When it does dry out it is usually badly rutted because of the fact that the drying began at the top, and in its earlier stages wheels would break through to the softer mud below. If the road has been improved with a hard surface, then the heave caused by the frost and the settlement caused by the thaw are likely to leave the surface cracked and broken and subject to early disintegration.

Many road builders consider that the best and cheapest method of subdrainage is by means of ordinary farm tile or pipe, made either of vitrified clay or of Portland cement concrete. The concrete pipe may be made on the ground if the materials are convenient. This course is followed by some quite prominent road officials and by many contractors.

The molds for concrete pipe are comparatively inexpensive, and the work of making them can be carried on under a shed or other shelter during rainy weather, or at other times of enforced idleness on the road construction work. Only the best Portland cement should be used and mixed with fine sand and gravel in the proportions of 1 part cement, 2 parts sand, and 3 parts gravel. Some authorities claim that there should be but 4 parts of sand and gravel combined, which proportion will certainly add to the strength of the pipe. The gravel must be clean and free from any possible coating of clay or other surface covering. It is better if the gravel be washed, and no individual pebble should exceed in diameter one-half the thickness of the wall of the pipe. Gravel which will pass through a $\frac{3}{4}$-inch mesh sieve is usually considered as large as it is advisable to use under any circumstances.

The concrete should be mixed wet enough so that the water will be forced out of the top of the mold and should be rammed into the mold until the water no longer appears at the top. The ramming, however, should be done quickly, so as to be completed before the concrete begins to set. Concrete pipe should season for at least two or three months before being used, and the temperature should not be allowed to get within less than 8 or 10 degrees above the freezing-point.

To determine what size of pipe to use requires the exercise of good common sense. There are but few data on the subject, by reason of the fact that after the pipe has been placed undergound there is no practical means of measuring the flow of water through it, or how much more or less would have flowed through a larger or smaller pipe. The nearest suggestion that can be given as a practical guide is the experience of farmers in the neighborhood where the road is to be subdrained, who have used pipe or tile in the same kind of soil and know whether the size they used was large enough to serve the purpose. The usual custom is 4 to 6 inches in road work.

All clay pipe should be thoroughly vitrified, but it is not essential that it be glazed. The sections are usually made 1 foot long for the smaller sizes and 2 feet for the larger sizes and with bell and spigot ends. Concrete pipe is made with square ends and of practically the same lengths, as compared with the size, as the vitrified clay pipe. The joints of the concrete pipe are usually wrapped in tar paper to keep the sediment from finding its way into the pipe, and also for maintaining a correct alignment of the pipe during the filling in process.

There are various other kinds of drains which are often used. One is a loose stone drain. In constructing this drain the trench is dug to the proper depth and grade, and about a foot wide at the bottom. Broken stone or gravel of coarse sizes, ranging from 1½ to 3 inches, is put in the bottom of the trench to a depth of about a foot; then a layer of smaller stone or gravel to prevent the earth filling from working down among the larger stones, and then the trench is filled. Another type of drain may be built of flat field stone by putting a layer along the bottom, then a row of stones along each side, and another layer of flat stones for the top. Sometimes small logs are used laid a few inches apart in the bottom of the trench and covered with flat stones.

None of these drains, however, seem to give the satisfactory results which are found by using pipe. They are much more likely to clog up and require expensive overhauling. Besides, there are liable to be differences in the character of the soils which would make such a ditch while perfectly good in one place, a source of danger from undermining or washing in another.

There are several ways of laying the subdrains, especially if they be of pipe. Probably the practice most generally employed is to place them underneath the side ditches, at a depth of about 3 feet below the bottom of the ditch. This depth is not arbitrary; it may be necessary to place it deeper in some soils and not quite so deep in others. Sometimes a single line of pipe is placed under the center of the roadway; but this method is not recommended except in rare instances, for the reason that any damage to the subdrain would damage the road and repairs could not be made without digging up the road surface. On heavy grades the subdrains should be laid under the shoulders of the road to avoid digging up the earth at the bottom of the side ditches, and so loosening it that it would wash out with a heavy rush of storm water.

Sometimes the subdrains are made so as to lead from the center of the road each way, extending diagonally down the grade and emptying into the side ditches. These transverse drains are also necessary in places where there is a springy or spongy spot of earth under the roadway.

The trench for a 5-inch pipe is usually made 12 inches wide at the bottom with sides slanting a trifle outward. In some cases a special tool is used to shape the center of the bottom of the ditch to fit the pipe. In other cases a layer of coarse sand or fine gravel is placed in the bottom of the trench in which the pipe may be bedded. Sometimes, but not often, a board is placed in the bottom of the trench and the pipe laid on it.

All subdrains should be carried to a proper outlet, usually connected with the surface drainage system. However, as the outflow of the subdrain is rarely heavy in volume it may often be carried into adjacent fields without causing much damage. The cutting the trench and laying the pipe should begin at the outlet end. If pipe with bell ends is used, the bell end should be upgrade. If the pipe is carried to an outlet through a bank, the outer end should be protected by concrete or masonry. Care must be taken in filling in the trench after the pipe is laid so as not to disturb its position. The filling may be of gravel, broken stone, or earth, according to the location and the necessity for a firm covering.

The size of the pipe bears a distinct relation to the

grade at which it is to be placed. Some road builders insist that a grade of at least 5 inches to the 100 feet is essential; others claim that 1 inch, or even $\frac{1}{2}$ inch is sufficient. The most popular practice seems to consider that 5 inches is little enough to provide a clear flow and avoid chances of settlement which might produce a low place somewhere. It must be considered that as the water finds its way into the pipe from the earth about it, if a sag developed at any point the water would be forced out at that point through the same openings at which it entered and saturate the base of the roadway.

In figuring the depth of a subdrain the soil must be considered first. In clay and other plastic soils a pipe or tile drain will drain an area about six times its depth on each side. Thus a pipe laid 3 feet deep should drain the subsurface water out of such soils 18 feet on each side, or 36 feet in all. In more porous soils the distance at which the earth is drained is much greater, sometimes more than ten or twelve times the depth of the ditch.

Experiments have shown that the earth-water does not drain into the subdrains until the water in the entire drained area reaches a level, but that the line of drainage is a practically straight line from the pipe to the surface of the earth at a given distance. For instance, if a certain soil is stated to drain 30 feet to a pipe 3 feet

(Courtesy U. S. Department of Agriculture.)

Fig. 10.—This is a photograph of a model made by the United States Office of Public Roads and Rural Engineering. It shows side-hill drainage, concrete culvert, surface ditch, side drain, drop inlet, culvert, V stone drain, side outlet, blind drain, center drain, laterals, and cobble gutters.

below the surface, the 30 feet will be the limit at which the drainage is apparent near the surface. At 20 feet from the pipe 1 foot of the earth will be drained, and at 10 feet there would be 2 feet of dry earth on the surface. Taking, then, a road 30 feet wide with a sub-drain pipe 3 feet under each side ditch, the subsurface water would be drained from the center of the road to a depth of $1\frac{1}{2}$ feet, in addition to the height of the crown of the road. Except in the most severe climates this depth should be sufficient to prevent frost-heaving to any appreciable extent.

CHAPTER V

ROAD FOUNDATIONS

THE natural earth is the final foundation on which all road structures must ultimately rest. Therefore, a study of road foundations is a study of the characteristics and conditions of the ground, and the methods by which a concentrated load on the road surface may be distributed over a ground area large enough to carry it without damage to the road structure.

The subject of road foundations is naturally classified into two parts: Natural Foundations and Artificial Foundations. The natural foundations are those which relate to all cases alike—the original earth base which must support the entire structure; the artificial foundations are those which support the various surfacings which modern traffic demands and convey the load pressure to the natural foundations. The Second International Road Congress which was held in Brussels in 1910 presented the following as one of its conclusions:

"The strength of road foundations should be increased in proportion as the supporting power of the ground decreases." In other words, the weaker the natural foundation may be, the stronger the artificial foundation must be made.

In figuring out the carrying capacity of the natural foundation there must be taken into consideration the soil, the subsoil, and sometimes the geological formation. The usual soils are gravel, sand, clay, shale, loam, and sometimes marl, peat, and muck. The subsoils usually consist of the same or similar materials, but often hardened into what is known as "hard-pan," and often the subsoil is almost or entirely without the element of decayed vegetation, while the surface soil may be rich with it.

Soils are formed by the decomposition or "rotting" of mineral, vegetable, and animal matter. By far the largest element is the mineral, and comes from the breaking down of various kinds of rock during the course of ages through wind, water, cold, heat, and, in the northern portion of the United States especially, through glacial action.

Scientists state that at some remote age in the past the northern portion of North America east of the Rocky Mountains was covered by a vast glacier, or series of glaciers, forced down from the north; that in its formation it absorbed vast quantities of gravel, rock, and other minerals which were carried with it until it finally rested, with its southern edge about the Ohio River, and approaching the same latitude east and west from there, not passing southward over the Ozark region in Missouri, and reaching the Atlantic a little south of the latitude of New York City. We are told that most, if not all, of the gravel and boulder deposits in this great section of country were carried by the glacier, and when it melted the courses of the rivers of the region were created and the hills and valleys established. It is a demonstrated fact that much of the hard gravel found in most of the Northern States is of the same geological formation as rocks which are found north of the Great Lakes.

Gravel consists of small pieces of rock which have become smooth and rounded by wearing against each other or against some other substance. It ranges in size upward from that of coarse sand, with which it is often found mixed, to stones of a size where they are called by other names, but with the same general characteristics. Cobblestones and boulders and various local names are given the larger stones in different localities. In the smaller or gravel size the pebbles are often found coated with a thin covering of clay, which would suggest that the water by which it was surrounded at the time of its deposit held a large amount of clay in solution.

While these glacial gravels are usually very hard, there are vast deposits of a softer gravel over many of the Southern States. The scientists have not yet stated authoritatively the origin of these gravels, which are found in beds on or near the surface. One name given to them is the "Lafayette" formation. They compact readily, make a good road foundation, and except for lack of hardness, to resist wear, would make an excellent road surface.

Sand is the name generally applied to any stony-like mineral substance which has, by decomposition or attrition, broken up or been worn into small particles. There are great variations in sands as to size of grain, shape of grain, and the mineral elements entering into their composition. Practically all sands have more or less quartz in their composition. Some sands are composed of particles which are round, or rounded, like a very fine gravel; others are what is known as "sharp" sand, in which the grains are angular. It is the sharp sand that is usually specified in all kinds of road work on account of its greater stability. Some sand packs hard when dry and becomes almost a quicksand when saturated with water; others, like the sand on many ocean beaches, are hard when wet, but are almost impassable for vehicles when dry. Even in walking over them the feet sink into some dry sands nearly to the shoe tops.

There are as wide differences in clays as in sands. Some clays will dissolve easily in water; others are soluble only after long immersion. Some will wash away easily and quickly, others seem to shed water almost as if oiled. Only experiments with the local clay in any section of country will furnish a safe guide for its use in road foundations. A clay subsoil under a marsh, however, can generally be depended on, as it has withstood the action of water and moisture long enough to demonstrate its reliability. Shale has nearly the same component parts as clay. It has evidently become hardened by being covered up with earth, and disintegrates rapidly on exposure to the air and heat and cold.

Loam is a mixture of almost all the various elements, but mostly clay and sand. It is found in various shades of color and in a variety of consistency when wet. A black loam along the Mississippi River bottoms, which is exceedingly sticky when wet, is known as "gumbo." Marl is a hard clayey substance which has not enough lime in it to cause it to harden into limestone. Exposure to the air has the effect of softening it.

Peat is vegetable matter which was decomposed under water. Muck is earth having a large proportion of decayed vegetable matter in its composition. Neither peat nor muck drains readily, but both dry rapidly when the water is removed from about them and they are exposed to the air and sun.

When it becomes necessary, by reason of any peculiarity of the soil or subsoil which will affect its value as a part of the road foundation, to have an analysis made to determine its usefulness, samples may be sent to the United States Office of Public Roads and Rural Engineering, Department of Agriculture, Washington, D. C. That office has an adequate equipment and a competent force of chemists who will make the examination and report without cost at the request of road officials. Such a report is very useful, as it shows to the local road official just how his local materials can best be taken advantage of.

It occasionally happens, though not often, that a section of road has a foundation of solid rock. Such instances usually occur in cuts or on side hills. On the rare occasions where this condition is found it is only necessary to prepare the side ditches carefully, so as to provide for the run-off of the water, and then install such artificial foundation for the surfacing material as the character of the surface may require.

The first requisite in the preparation of a natural foundation on any kind of a subsoil is to see that it is well drained; and that the drains will carry off the natural rainfall, the excessive storm water, and also the water from the subdrainage channels, without permitting water to settle on or about the road, or the fill, or the foundations.

In sand or clay or loam soils the natural earth should be disturbed as little as possible in preparing the natural foundations, especially where the factor of subdrainage does not require it. Experience has shown that much is gained by utilizing the strength of the earth in its original position.

In muck, quicksand, or other soft earth judgment must be exercised. Sometimes it becomes necessary to dig the muck out and make a fill of stone, brick-bats, gravel, cinders, or of hard earth. Previous to this, however, test holes should be dug at occasional intervals to determine the depth of the muck and its character and solidity further down. In some instances large flat stones have been used as a base to prevent the road from settling into the muck.

The variety of conditions encountered where roads cross low, soft places makes necessary more study of this phase of natural foundations than the length of the roads over them would suggest. But such conditions arise frequently and must be considered carefully.

Occasionally, for a short distance, a condition is found where the soft mud or muck seems to have no bottom; where load after load of earth and stone and other materials have been dumped to create a roadway, only to find that the settling goes on regularly, and under some stress of heavy rain or a melting snow freshet the entire road may sink out of sight. For this condition the first requisite is a careful investigation to determine the depth of the soft deposit, which is often muck mixed with a quicksand. This may be done by building a wooden platform, supplied with windlasses, and sinking an iron rod or iron piping until the bottom is reached.

Then it may be necessary to drive a row of piles along each side of the road, the piles being close together and long enough so that they get a good holding in the solid earth at the bottom of the mud. The tops of the piles may be held firmly together by planks spiked to them. Then the filling may proceed between the two rows of piles with any material available until a solid embankment is made.

An unusual treatment of such a condition is sometimes reported. In one instance a marsh was encountered which had given much trouble. It was about $\frac{1}{3}$ mile across, and the soft mud was so deep that any piles that would reach to the bottom would be expensive and difficult to obtain. In the neighborhood was an extensive growth of saplings, ranging from 20 to 40 feet high and up to 4 inches in diameter at the butt.

The official in charge built a foundation the whole length of the road over the marsh by placing these saplings in interlaced courses. The foundation was built 40 feet wide. The first course of saplings was laid crosswise of the road with the butts outward, 6 or 8 inches thick. Then a course was laid lengthwise of the road, but of slightly smaller saplings, with the butts thrown forward to the marsh and the branches on top of the cross-layer, and the next section of those placed crosswise holding down the butts and branches of those laid lengthwise. Then a couple more courses were put on until the "mat" of saplings, thoroughly interlaced, formed a body approximately 3 feet thick from one bank to the other. This was then covered with about $2\frac{1}{2}$ feet of coarse gravel and opened to traffic. It was not considered wise at that time to compact the gravel with a heavy roller, but as no considerable settlement was observed by the next spring, the surface was trimmed with a road machine, rolled, a surface of about 3 inches of gravel ranging up to $\frac{3}{4}$ inch in size placed on it, and then rolled with a 10-ton roller to a finished surface.

Under a heavy traffic, as traffic is measured on country roads, 250 to 500 vehicles a day of all classes, this road has shown no material settlement nor any but surface deterioration during the three years it has been in use.

In another instance a careful survey of levels indicated that an entire swamp, over which the road must pass, could be partially drained by blasting out a channel of less than 100 feet at the outlet of the swamp more than half a mile from the road. This was done, and during ten months of the year the swamp was comparatively dry; the water level having been lowered nearly 3 feet. During the dry period the road foundation, made of stone and earth and gravel from higher ground adjacent, became solid enough so that it was not affected by the water which covered the swamp during the month or two of the wet season. It may be worth while to state, though the fact has no relation to road foundations, that the owners of that particular swamp have formed an organization, and raised the money to deepen the outlet and reclaim the several hundred acres of swamp lands involved, so as to make them tillable.

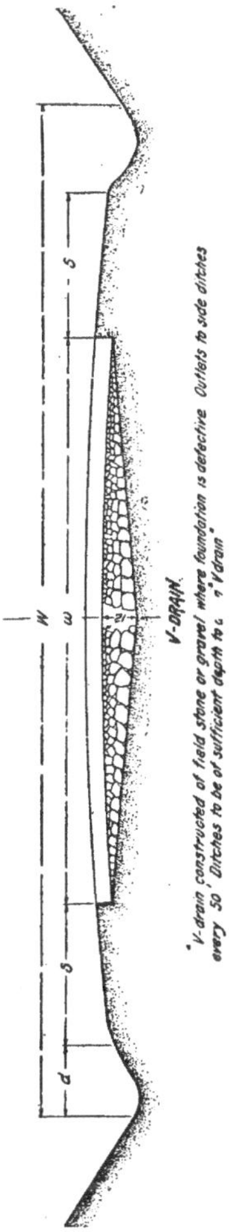

Fig. 11.—Stone foundation forming a V-drain at center.

In some sections of the country the natural foundation has been used to some extent as the immediate foundation for the surfacing material. Stone blocks, brick blocks, and other surfaces have been laid directly on the subsoil, with perhaps only a thin layer of sand, and have worn for many years. Brick has been laid on a sand subsoil, and in some instances the roads have held their shape for several years. The percentage of failures of these methods, however, has been so large that the methods are not recommended. A new type of construction, known as "sand asphalt," has been laid directly on the subsoil. These roads have not been down long enough to determine either their lasting qualities or whether the natural soil furnishes an adequate foundation. These roads are treated under their appropriate heads in other chapters.

Artificial foundations vary with the conditions of the subsoil and the types of surfacing. For macadam roads a foundation of broken stone is laid directly on the subgrade, which has been properly shaped and rolled. Care must be taken in preparing the subgrade that no soft spots remain; when such spots appear under the roller the soft earth should be removed, and earth which will compact firmly put in its place.

The most approved practice calls for broken stone ranging in size from $1\frac{1}{2}$ to 3 inches. This is usually laid to a depth of 5 or 6 inches after being rolled, but in determining the thickness of the foundation consideration must be given to the amount and character of the traffic the road is expected to carry. On roads where much of the traffic is composed of heavy trucks 8 or 9 inches of stone may be necessary; where the traffic consists mostly of light pleasure vehicles 4 inches of stone have been found adequate for the foundation.

A practical illustration of this particular feature is found in the driveways of the Gettysburg Battlefield, many miles in extent. A few of the roads, those forming the main thoroughfares through the Federal reservation, are constructed with foundations 6 to 8 inches in thickness; but by far the greater part of the road mileage is built on 4-inch foundations.

For more than a decade these roads have stood up under a very heavy travel of pleasure vehicles and have given satisfaction to great numbers of visitors and tourists. Only pleasure traffic is allowed except on those main thoroughfares which were formerly county roads, unless permit is given by the Federal authorities.

It has seemed necessary that such permits be issued in several instances. Usually the necessity arose for the purpose of conveying a heavy stone or granite base of some monument to the location which had been selected for its erection. There are many of these monuments in various places on the battlefield, marking historic spots or representing the patriotism of other sections of the country.

The bases for these monuments are usually solid blocks of stone or granite, weighing sometimes 10 to 20 tons. To get them to the proper location they must be hauled on trucks from the railroad station in the village of Gettysburg to the site selected over the roads most available. These roads are designated by the authorities when the permit is given, and provision is always made for as large a proportion of the distance as possible over the main roads.

It is the almost invariable experience that the main roads hold up, except perhaps in spots, under these loads. Almost invariably, also, it is found that when the driveways with the lighter foundation are reached they are crushed and displaced, requiring practical reconstruction of the section of roadway traversed.

But it has also been found that as a matter of economy it is cheaper to reconstruct the roads in these occasional instances than it would be to have made the foundations heavy over the entire system. These roads cost approximately $8000 a mile, with 4 inches of foundation and 3 inches of surfacing. Four inches more foundation would probably have cost $3500 or $4000 a mile additional. The interest on the money thus saved would pay for many times the repairs which have been necessary.

In preparing the subgrade for the stone the usual custom in the United States is to bring it to the same rounded shape as the surface of the road, so that the foundation course of stone shall be uniform in thickness under the road for its full width. Some road builders, however, prefer to make the subgrade flat, and to make the rounded surface by putting more stone in the middle and less on the sides. In a 5-inch foundation this plan would call for 6 inches of stone in the center and 4 inches at the sides, giving the road a crown of 2 inches. While this method has been used much abroad, especially on English roads with Telford foundations, American practice only seems to favor it in exceptional conditions.

One of the arguments favoring the rounded rather than the flat subgrade is that the present-day traffic is so different from that of former years, especially in the weight of vehicles, that a flat subgrade under present conditions might soon become settled in the middle, so as to be lower at the center than at the sides. In such cases moisture might settle under the middle of the roadway instead of draining away at the sides, and the frost action would prove damaging. The success of the flat subgrade abroad, it is argued, is due to the fact that for scores of years the stone has been compacted by an ordinary traffic no part of which was heavy enough to displace the subgrade, until the stone foundations have become so thick and so dense that they can be depended on to carry almost any load without affecting the subgrade.

Telford foundations are used where the subsoil is soft, so that broken stone would be likely to sink in. The Telford foundation takes its name from its inventor, who was a noted road builder in the south of England a century ago.

In putting in a Telford foundation the subgrade is usually shaped with a crown the same as the crown of the finished road. Stones are then set up edgewise with the widest and flattest edge down, and with the longest dimension crosswise the road, and the points sticking up. The stones are placed as close together as possible. The top points are then broken off with a sledge and the pieces driven down as wedges betwen the stones. This breaking off the tops of the stones brings them to a uniform height and the wedging makes them solid. The foundation should then be rolled with a heavy roller until it is hard and solid. Then it is ready for whatever surfacing has been selected.

It is proper to state in this connection that Telford completed his roads by the addition of several inches of broken stone. Modern practice, however, makes greater use of his foundation than of his completed road.

There is considerable variation in the sizes of stone employed in building Telford foundations in different states. Connecticut and New Jersey require a depth of 8 inches; New York, 6 to 8 inches; Massachusetts, 5 to 6 inches. The width of the stones as set may reach anywhere from 6 to 10 inches in Connecticut; 4 to 10 inches in Massachusetts and New York, while New Jersey uses smaller stones not exceeding 4 inches in width. In length, placed crosswise of the road, Connecticut uses stone 8 to 18 inches long; Massachusetts and New York, 6 to 15 inches; and New Jersey, any available length up to 10 inches. In Massachusetts, however, 2 inches of gravel is generally rolled into the subgrade before the stones are placed.

Concrete foundations are necessary where brick or other blocks are to be used for the surface, and sometimes when other surfaces are employed. If the traffic over the road is to consist mostly of heavy loads, concrete foundations should be placed under most kinds of bituminous surfaces. The thickness of the concrete foundation must depend on the nature of the subsoil, and on the prospective loads to be carried, taken in connection with each other. Many road authorities claim that the concrete should not be less than 5 or 6 inches thick; on the other hand, the state of California, after careful investigation and tests, is building more than 1000 miles of state roads on a 4-inch concrete foundation.

The concrete for road foundations is made either wet or dry, according to circumstances. If water is plenty most road builders prefer a wet mix; but there should not be water enough in it so that there will be any material run-off. A dry mix requires more tamping so as to get the moisture thoroughly distributed, and it should never be so dry that it cannot be tamped into a solid mass. At its best, dry concrete is not so dense as that which is placed wet. A concrete which is a medium between a dry and wet mix is used by many road builders. This is of the consistency of a stiff mortar, just soft enough to be tamped into place before it sets.

On large contracts, or where there is much work to be done, it is usually considered more economical to use mixing machines. Of these there is a variety, and each mixing machine possesses some peculiarity which makes it particularly available for some especial work under certain conditions. As between "batch" mixers and "continuous" mixers a majority of engineers seem to prefer the "batch" class; while a large number of practical concrete men prefer the "continuous" machines. These engineers claim that the batch mixer is more nearly "fool-proof," insures a more uniform and thorough mix. Some practical men claim that the mixing can be done just as well and much faster with a continuous machine if properly handled; and that no one has any business attempting to mix concrete who does not know how or does not know when his product is in the best possible condition. Most of the modern mixing machines of either class have devices for depositing the mixed concrete in its proper place as a part of the labor-saving proposition.

Hand mixing is done on a mixing platform made of planks spiked close together on cross-pieces, either of plank or scantling. The sand and cement are first thoroughly mixed by turning with shovels, beginning at the sides and working toward the center; then reversing and turning toward the center. When the cement and sand are thoroughly mixed the stone or gravel aggregate is added and the turning is continued until all the materials are thoroughly mixed in a dry state, and then the water is added.

If plenty of wheelbarrows are at hand the water may be put on the entire batch on the mixing platform; the turning and watering and mixing may proceed until the concrete mass is properly mixed, and be loaded at once on the barrows and deposited in place, raked into position, and tamped until the proper density is secured. If there are not enough barrows a part of the batch may be watered and mixed at a time, and another lot mixed while the first is being taken away, all depending on the size of the batch and the rapidity of removal. Care must be taken in such cases to see that there is no wait long enough for the first set of the concrete to take place before the next delivery.

Concrete in road foundations should not be laid when the temperature is within 10 degrees of the freezing-point. At night, in any weather, the end of the work should be covered with tarpaulins and kept wet. The foundation itself after it has set should be covered with sand or dirt to a depth of 2 or 3 inches and kept moist for a period ranging from three days to a week, so that the final set may take place without damage. Some road builders make a practice of setting a board up edgewise at the end of each day's work and making an expansion joint. While many good authorities hold that expansion joints are not necessary in foundations, others hold that it is better to provide them than to run the risk of cracks where one day's fresh concrete meets that which was placed the day before.

The mineral aggregate of the concrete for a road foundation should be of broken stone or gravel, clean and free from dirt, and especially free from clay. Old stone, which has been used in macadam roads or otherwise, does not make a satisfactory aggregate for concrete. Fresh stone with clean sides, showing fresh breaking, is best. In the use of gravel, if the gravel be not perfectly clean, as in most cases of bank gravel, it should be washed. Creek or river gravel does not usually need washing; but care must be taken to see that it is well graded; that is, that there is a proper proportion of different sized stones.

Stone or gravel as the aggregate for a road foundation does not require the hardness or strength that is necessary in road surfaces, especially if the subgrade be of reasonably firm earth. Almost any available material, properly prepared, will answer the purpose. There are patented preparations by means of which foundations have been made of the natural soil, loam, sand, etc., and Portland cement. With the proper chemical proportions there seems no reason why such mixtures should not prove permanent.

(Detailed information on this subject may be secured from the United States Office of Public Roads and Rural Engineering, Department of Agriculture, Washington, D. C.)

Concrete foundations should never be placed on an old macadam or gravel base, which has been built up and worn down by years of travel. The reason for this is that it is practically impossible to get the concrete top and the old base joined firmly enough so that water will not get in between; then the frost breaks up the concrete. Besides, the old macadam or gravel base is usually fully as solid and firm as new concrete would be. Repair of defects or correction of alignment may be made in the usual way, by roughening the surface, and using fresh stone with a cement grout or soft mortar. Sometimes an entire old macadam or gravel foundation, after having been scarified, repaired, and rolled, may be made uniform and solid by an application of a cement grout, probably a 1 : 1 or a 1 : 1½ mix, which will fill all the voids or interstices and make a thoroughly solid foundation.

Much, however, depends on the surfacing material which is to be placed on the foundation. These are treated in the Second Part, under the various headings of "Roads." Generally speaking, old stone or gravel foundations are more economically used when employed as bases for bituminous surfaces. One reason for this is that the bituminous surface fits closely into the small irregularities of the foundation, and leaves no space for water to settle; second, it is more nearly waterproof than other surfaces; and third, that the expansion and contraction due to heat and cold does not pull it loose from the foundation.

Notwithstanding the theories of some engineers, there is no economic sense in digging up or destroying an old gravel or macadam foundation. Such a foundation, settled and compacted by years of traffic, represents a certain definite investment for the community; and its destruction is a waste of community funds. In such cases the study should be for the surface best adapted to the existing foundation, as the foundation should represent at least half, usually more, of the value of the completed road.

In some cases where the subgrade presented irregularities and unevenness as to density, concrete slabs have been used as foundations. Sometimes these slabs have been reinforced by expanded metal or heavy wire mesh. The slabs range from 12 to 20 feet in dimensions and are about 6 inches thick. The subgrade is prepared as carefully as possible, the slabs set close together, and the joints filled with cement grout. Usually a sand cushion of 3 or 4 inches is placed on the subgrade under the concrete slabs.

A paving company manufacturing a patented pavement has at times employed its surfacing methods in constructing foundations. This method, which is understood not to be patented as to foundations, consists of placing broken stone or gravel to the proper depth and pouring grout over it while it is being rolled. The grouting and rolling continues until all the voids are supposed to be filled, and then the work is left to set in the usual way. Especial care must be exercised in using this method to stop the rolling before the first set of the cement has begun; otherwise the strength and holding power of the cement will be interfered with.

Foundations of bituminous concrete, using asphalt or tar instead of cement, with broken stone or gravel, have been laid. Reports show that while serviceable, they have not the strength of Portland cement concrete, and that the only special reason for their use would be their cheapness in some given locality. In any remote place where cement is especially expensive by reason of freights or otherwise, and tar or asphalt cheap because of local production, the bituminous materials can be depended on to make a foundation which will last for many years under proper conditions.

CHAPTER VI

ROAD SURFACES

THE surface of a road must be considered in connection with the grade; with the smoothness; with the traction power required to pull loads; with the use of the road either for pleasure vehicles or for heavy truck traffic; and with its permanency or "staying" qualities under different conditions of traffic and climate.

The surfaces of standard types of roads are described under the various chapters relating to the construction of those roads. Modern practice, however, has a tendency to depart from types to the extent of putting almost any kind of a surface on almost any kind of a foundation in the effort to get the best road for the least money. Sometimes experiments along this line have been very successful; in other instances they have been wasteful—if not total failures.

As to grade and traction power. As automobile trucks are so constructed as to go over any ordinary grade, horse-power must be the basis for figuring. According to the conclusions which have been reached by engineers who have studied and experimented on the subject, a horse will draw four times as much load-weight on a level as it can draw up a 10 per cent. grade. For instance, if a team of horses can draw a load of 3000 pounds on a level road, the same amount of pulling would draw 1500 pounds on a grade ranging from 5 to 6 per cent., and 750 pounds on a 10 per cent. grade.

Or, to put it the other way: If two horses can pull 3000 pounds on a level, it will require four horses to pull it up a 5 per cent. grade, and eight horses to pull it up a 10 per cent. grade.

The amount that a horse can pull, of course, on a level or any grade depends on the smoothness or lack of smoothness of the surface of the road. If the surface of the road be hard and smooth, theoretically it would require as much power to haul a load up a 10 per cent. grade as the combined power required to haul it 100 feet on a level and to lift it 10 feet. If the surface be rough the additional power required must be gauged by the degree of roughness.

It is usually stated that an average draft horse exercises a pulling power of about 180 pounds, and that this can be increased at times, and for brief periods, to 500 pounds. These periods of special pull are exercised in starting the load and may be employed in such emergencies as make extra pulls necessary; as in defective spots or holes in the road; or an especially steep short hill which can be covered in a few minutes by putting forth the extra energy of which the horse is temporarily capable. The physical proposition might be compared to that of a man, who might be unable to carry a weight of 50 pounds for eight hours, but could carry 200 pounds for fifteen minutes.

In connection with the pulling power of horses must be considered not only the easy movement of the load on wheels, but the foothold of the horses. Where the surface is smoothest the power required to move the load is naturally least. At the same time the foothold of the horse becomes less as the surface becomes smooth. On an earth or gravel road the horse may have a better foothold, and therefore a greater pulling power; while on a smooth, hard surface, such as concrete or asphalt, the foothold will not be so firm, and the power necessary to pull the load will be correspondingly less.

Up to a certain grade, however, the authorities all favor the smooth road; but as to just where the smooth surface becomes a disadvantage and the rough surface an advantage there is much difference of opinion. Some assert that 7 or 8 per cent. grades are as steep as smooth hard surfaces should be built; others claim that a rough surface is more economical from a 10 per cent. grade up; while in various localities throughout the country, notably in Allegheny County, Pennsylvania, hard smooth surfaces are used on grades as steep as 14 per cent.

One fact may be noted. When smooth, hard roads

are built in country districts it requires a little time for farm horses to get accustomed to them so that they can exercise their full draft power. With care in driving, however, they soon adjust themselves to the new conditions, and become as sure footed as horses trained to city pavements.

The Massachusetts Highway Department has made many experiments in surfacing roads on various kinds of foundations and in maintaining the surfaces in good travelable condition. The reports of these experiments are voluminous and range from official reports to papers and addresses delivered before scientific and other conventions and published in magazines and official proceedings. They range all the way from the application of oil on a graded earth road to patching cracked concrete with tar or asphalt.

While each particular road presents a different condition, the approved Massachusetts plan for creating a wearing surface on a road, whether it be an earth road, a gravel road, or a macadam road, is to make such an application of asphaltic material, with a sand, or fine gravel, or stone-chip covering, as will produce a mat or carpet which will take the wear, shed the water, and protect the road from damage. This surfacing is patched or renewed whenever it is necessary to maintain the road. Traffic conditions in that state require that the roads be kept in good condition.

8

The asphaltic material used varies. In some instances light asphaltic oil—45 per cent. asphalt—is first used at the rate of $\frac{1}{3}$ to 1 gallon per square yard. That is to get the proper penetration. Afterward more light oil may be used in smaller quantity, or a heavy oil —60 to 65 per cent.—may be added, and the surface covered with fine gravel, sand, or stone chips. After the surface has been established, one application of the heavy oil each year is sufficient in most cases; but, of course, the amount, and the grade of the asphaltic oil, and the frequency of its application are matters which depend entirely on local conditions of climate, soil, foundation, and traffic.

In maintaining these surfaces the automobile with pneumatic tires is one of the greatest factors to be considered. Trucks with steel tires easily damage this road surface if there are no automobiles. Motor trucks with solid tires do less damage; but the pneumatic automobile tires are a positive advantage to the road, ironing and smoothing out the surface which may have been cut up by other vehicles. It has been stated that it is cheaper to maintain a perfect surface of this character under a travel of a thousand automobiles a day than under that of fifty loaded wagons with steel tires. On the other hand, the fifty loaded wagons will do little damage if three hundred to five hundred automobiles with pneumatic tires are going over the road at the same time, as they smooth out most of the irregularities caused by the heavy loads on the steel tires.

But with an earth base, especially if the earth be of a soft character, easily permeable by water, the surfacing is likely to be cut through and rutted or holes developed at any time a heavy rain occurs. The asphaltic top only protects the road underneath to the extent that it makes it smooth and waterproof under ordinary conditions. When the conditions change, other or further methods are necessary.

When a road surface has been treated with asphaltic oil, especially with heavy oil, the surface particles become coated with the asphalt. If the surface becomes broken up and rutted by the travel, the road can be shaped again by the road machine or the road drag, and the particles will amalgamate together and the surface will resume its former condition, especially if there have been several applications so that the surface material has been thoroughly saturated with the oil. It must always be remembered, however, that when water gets under the surface, and the traffic breaks through, it requires but few vehicles passing over the spot to make a mud-hole; and a mud-hole or other break in the surface must have prompt attention to prevent its expansion to a really "bad place in the road."

It is a general impression, under average conditions, that with a fairly good foundation it is more economical to maintain a good surface than to let the original surface wear out and build a new one. In many instances where the traffic has increased from an average of fifty vehicles a day of all kinds to more than three hundred a day—the increase being mostly in motor vehicles—good wearing surfaces have been maintained at a comparatively small expense per year for ten or twelve years. As every road surface, new or old, needs constant attention, the question of cost of upkeep must be the difference between the cost of repairs on a new road and on an old one.

A 15-foot road has 8800 square yards of surface to the mile. Unless the traffic be very heavy, or unless the traffic changes in character rapidly, a new hard surfaced road should be kept in good condition for the first three to five years at an expense of not more than 2 to 5 cents a square yard; that is, from $176 to $440 per mile. The experiences of various highway departments have shown that the cost of repairs gradually increases as the road grows older. These increases in cost, however, are invariably shown to be due to the fact that there has been a very great increase in the traffic. In this connection it may be well to state that according to the most careful estimates the vehicles traveling over the roads of the United States considerably more than doubled between the years 1910 and 1915.

It has been figured out quite carefully that when the

cost of keeping a "mat" or "carpet" surface—as previously described—in good condition exceeds 10 or 12 cents a square yard per year, it is more economical to build a new road of a character and type which will stand the traffic. The basis of the calculation is the cost of making the new road, or such part of it as must be rebuilt. If the yearly interest on the cost of reconstruction be less than the annual cost of maintaining the old road, it is more economical to rebuild the road. This is especially true by reason of the fact that the new road should be built to withstand the stress of present and future travel, the constant growth of which would make the repairs of the old road constantly and rapidly increase in cost.

The United States Office of Public Roads began about 1908 or 1909 to make experiments in regard to various materials and methods of road surfacing, both on its own account and in connection with other interests. These experiments relate both to dust suppression and road preservation. They were made in various parts of the United States and under greatly varying conditions, and the various bulletins issued by the Department of Agriculture, which are obtainable by any road official, give in detail the character and location of the experiments and the successes or failures which resulted.

It must be borne in mind that until 1907 or 1908

the macadam or broken stone road (described in another chapter) was considered the highest and most nearly permanent type of construction applicable to country roads. Its successful use for one hundred and fifty years in France, and for nearly one hundred years in England, and for only triflingly shorter periods in the German and Scandinavian countries, and the employment of the same principle in the United States on the most highly improved roads, had established that form of construction as standard.

The advent and rapid development of automobile travel between the years 1900 and 1906 seriously interfered with the road surfaces. It was found that the previous methods of maintenance would not keep the roads in good condition. The first official notice of this deterioration of the roads under automobile traffic in this country was in the Annual Report of the Massachusetts Highway Department in 1907. Although the damage had been noted in other states, it had not been made a matter of official report.

So universal throughout the civilized world was the condition which threatened the destruction of broken stone roads that an International Road Congress was held in Paris in 1908 to determine what measures were necessary to preserve the roads, and the wisdom of the highway engineres of all civilization was brought to bear on the subject. Another International Road

Congress was held in Brussels in 1910, and the third in London in 1913.

In the earlier stages of the extraordinary wear on the roads the damage was attributed entirely to automobiles, largely because it had not existed prior to the use of automobiles. Study of the subject, however, demonstrated that it was the combined effect of steel-tired and pneumatic-tired vehicles which produced the destruction. The excellence of the surface of a broken stone road is due to the fact that the steel tires and steel-shod hoofs, while grinding some dust and small particles from the stone, packs that dust and fine particles down between the stones, forming at once a binder for the surfacing stone and a coating which prevents rapid wear. The rains, by washing the dust further down among the stones, aided in forming a hard, smooth surfaced, substantial road.

Then the automobile came with its pneumatic tires and low, rapidly moving body. The suction created by the passing of the machine over the road pulled the dust out of the crevices between the stones and threw it into the air to be blown away by the winds. Then horses and steel tires would break down more of the surfacing stone, the particles to be blown or thrown off as before. This alternate breaking and throwing off rapidly destroyed the best stone road surfaces.

For a time much attention was given to dust suppression as a cure for the evil. A number of materials were invented or developed for keeping down the dust which have been more or less successful, according to the intelligence with which they have been applied. In a large number of cases they have been very successful, especially on roads where the traffic did not increase to an extent beyond their usefulness.

It was soon found, however, that the suppression of the dust—keeping it down so that it would not fly in the air—did not produce satisfactory results on heavily traveled roads. Dust continued to form under the traffic, and though so loaded with tar or oil or chemicals that it would not fly in the air, it was unpleasant to travel over, and formed a very offensive mud after rains.

Then the experimentation turned to road binders and preservatives, so as to develop or discover methods of constructing road surfaces so that they would wear well; would produce but a small amount of dust; and as far as possible pack the dust again into the surface as an additional prevention against wear. Another class of experiments have been along lines of creating a surface so hard that the wear will be negligible, as is claimed for various modifications of Portland cement concrete. Several of the surfaces thus invented or developed have been patented and are known under trade names which can be found in any current copy of journals or magazines devoted to road improvement. Usually, each of such materials as are used and the methods of application may be depended on to produce satisfactory results under conditions favorable to their employment. It will be wise, however, for a road official to carefully study the relative conditions where any such surface has been successful, and those of the road under his control which he desires to improve.

Resiliency in a road surface has for many years been considered necessary, and still is so considered in many sections of the country. Resiliency is that quality in a road surface which is the opposite from rigid. A resilient surface "gives" to some extent under the impact of traffic. All bituminous surfaces are resilient; as are also, though to a less extent, brick and other blocks laid on a sand cushion. Concrete is rigid and presents an absolute resistance to the impact of traffic.

The theory is that a rigid road surface is injurious to horses and that it causes greater wear on automobile tires. This theory is disputed by advocates of concrete road surfaces, who contend that the horse becomes accustomed to a rigid surface as readily as to any other hard surface, and that the effect on rubber tires is favorable rather than otherwise. There is room for discussion on both sides of the question. It appears to be a fact that such popularity as rigid surfaces have achieved has been in the localities where motor traffic predominates; and that horse owners, almost without exception, prefer road surfaces in which there is some resiliency or "give" when the horses' feet strike it.

Road builders must keep in mind the fact that the same materials and methods of application produce different results in different sections of the country. A certain patented surface which involved a mixture of bituminous materials with a concrete surfacing was an entire success in one section of Illinois, and a positive failure in a locality where it was tried in Ohio. The same pavement, while maintaining its strength on one of the drives of Central Park in New York City, became so unsightly, by reason of the surface "peeling" in spots, that it was covered with asphalt by the park authorities. The patented road made an excellent foundation for the well-tried surfacing, but the combination was somewhat expensive.

One eminent engineer has asserted that water, properly applied, is the cheapest and best method of keeping down the dust and protecting the road surface. Possibly his views may be modified by a consideration of the cost of getting the water to the road. His view, as expressed, is that applications of water, applied with sprinklers during a season, in such quantities and at such intervals of time—whether once a day or five times a day, as shall keep the road moist, neither dusty nor muddy—can be made at less cost than applications of bituminous or other chemical materials which will produce an equally satisfactory result.

Experience does not bear out this theory in the opinion of most road officials who have tried it, except in those cases where water is unusually convenient, so that sprinkling wagons can be loaded at practically any point along the road and with practically no expense. Such a condition is very unusual. In most cases where the road surface suffers most from the creation of dust an ample water-supply is difficult or expensive to reach.

On roads of ordinary traffic, not exceeding two hundred to two hundred and fifty vehicles a day of all classes, and without especially heavy trucks, oiled or tarred surfaces have been applied frequently within recent years, and when properly applied, with a fair measure of success. In almost numberless instances the application has been made by persons either without knowledge, or with fragmentary information, of the proper methods. These have resulted in dissatisfaction and unpleasant experiences.

Oil or tar should not be placed on a road until the surface has been swept clear of dust. To put bituminous materials on top of a coat of dust is to create a condition where the wheels of vehicles will pick up the bituminized surface, leaving the roadway bare in spots and clogging the wheels of the vehicles; and footsteps of persons or animals will be tracked to the sidewalks and to adjacent buildings, carrying the marks of the bituminous material.

When such a surface is to be applied, the road should be swept clean of dust first with reed brooms, and then, if it be a well-bound stone road, with ordinary house brooms. Then the oil or tar may be applied, but always when the road is entirely dry.

There are several methods of application. A dozen or more different machines have been designed for this work, some of which are motor driven, some horse-drawn, and some pulled by hand. Some simply deposit the material on the roadway; others distribute it under pressure. Some are arranged to heat the material; others for delivery cold. The availability of either method depends on local conditions.

Some of the bituminous materials, whether oil or tar, when rightly applied and covered with a coating of stone chips, gravel, or coarse sand, may be opened for traffic within a few hours—perhaps the next day. This is especially the case where the application has been of a light oil, or an oil which has been applied hot, and in small quantity per square yard.

According to some road builders, to secure the best results, a light oil should be spread on the dry, swept, surface of the road, using about $\frac{1}{4}$ to $\frac{1}{3}$ gallon per square yard. This application should be left for twenty-four or forty-eight hours without covering, and without traffic, to allow the oil to sink well into the road surface. Then a similar amount of heavy oil should be applied, and the covering of sand or stone chips put on. If the heavy oil is applied hot, about 212° F., the road may be opened for traffic after six or eight hours. If the application is of heavy oil cold, at least twenty-four hours should elapse before traffic is permitted.

Some materials, especially some classes of tars, require that the road be kept closed to traffic for a week or more to give the material time to "set"; that is, to become hard enough so that it will not be picked up from the road by the traffic, and that it will not be tracked to adjacent walks and buildings.

When oil is to be heated it is usually shipped in tank cars which have steam radiating pipes, so that attachment may be made with a steam roller or any other steam-producing machine. Sometimes a car can be switched near enough to some local plant operated by steam power so that a steam hose may be attached to the radiators of the car. If the material is sufficiently heated in the car, and the distribution be prompt, it will probably not require further heating. However, there are a number of distributing machines on the market which have heating apparatus as a part of their construction. When shipped in barrels, the material may be heated in wheeled kettles at the roadside.

An eminent authority on highways (Professor Arthur H. Blanchard, of Columbia University) presents the following table of costs of what he considers an average condition: The road was treated with ½ gallon of heavy asphaltic oil, in two applications of ¼ gallon each, per square yard. The average haul was 2 miles for the oil and 2½ miles for the sand. No allowance is made for use of machinery or for profits to contractor, the work having been done on force account. The detailed costs per square yard are given as follows:

Cleaning and sweeping	$0.0056
Patching old surface	.0016
Cost of oil	.0319
Heating oil	.0031
Delivering oil	.0038
Distributing oil	.0029
Furnishing sand at side of road	.0165
Spreading sand	.0073
Watering	.0012
Rolling	.0002
Supervision	.0025
Total	$0.0766

It will be noted that nearly one-half the cost of the surfacing was the cost of the oil. The price of labor was $1.75 per day of eight hours.

Some experiments have been made in which the sand was heated; but the reports do not seem to show a sufficient degree of additional excellence to recommend the extra expense.

With ordinary traffic a road such as the one described should require but a single application of $\frac{1}{4}$ to $\frac{1}{3}$ gallon of heavy oil each subsequent year, with the usual sand covering, to keep it in good condition indefinitely.

CHAPTER VII

ROAD BRIDGES AND CULVERTS

THE greatest amount of water ever likely to pass under a bridge or through a culvert must be the measure of its size. On large streams this may be figured from past high-water records; on medium and smaller streams the number of acres or square miles drained, and the heaviest known rainfall or the heaviest snowfall followed by quick melting may be figured out to determine the greatest flow. If there is a railroad bridge over the same watercourse near by, it is probable that the engineering department of the railroad has a carefully worked out estimate of the stream flow. Nearly all railroad companies furnish these figures to highway officials on request. They are obtained by a survey of the drainage area. The heaviest rainfall and the steepness of the slopes are the other principal factors. In some cases springs and flowing wells are numerous enough to be taken into consideration.

It is most important from every standpoint that the waterway be large enough to carry off the water without obstruction. Thousands of bridges have been washed away and an almost limitless number of culverts washed out by the neglect of this primary principle.

In addition to securing data covering the extreme high water and low water for a long period of years, allowance must be made for the obstruction by piers and abutments, especially when a number of short spans are to be built across a wide channel. Most bridge experts and engineers hold that even though a considerable amount be added to the expense of the bridge, it should be built with water-way openings enough and large enough so that there shall be no holding back of water, but that the run-off must be perfectly free.

Culverts are generally divided into three classes: the pipe, box, and arch. Pipe culverts include those constructed of vitrified clay, cast iron, corrugated metal, and sometimes of concrete. Box culverts are built of concrete, stone, and wood. Arch culverts are usually built of concrete, reinforced or otherwise, stone, and brick. The selection of the type of culvert for any given locality should be based on the availability of materials and the economy of construction, although the size of the opening and the depth below the road surface are factors which must be considered.

The culvert, in addition to its function in carrying off water, must sustain the weight of the roadway above it and of the loads which go over the road. When the culvert is at a considerable distance below the road surface, say 4 or 6 feet or more, the pressure on it from loads is more evenly distributed than where it is within a foot or two of the surface. When it extends near to the road surface the culvert is likely to get the direct pressure of the load when the wheels pass over it, and consequently it must be of sufficient strength to carry the load without breaking or collapsing. In the employment of pipe culverts, where the embankment is not high enough to properly protect a large pipe, many road officials have put in a row of two or three or more small pipes side by side, either close together or separated a few inches, according to circumstances. For instance, three 12-inch pipes will usually answer all the water-carrying purposes of one 24-inch pipe, and will give an additional foot of embankment over them which, if the pipes be separated somewhat, will materially reduce the pressure from heavy loads.

Except in occasional instances it is hardly necessary to go into a detailed investigation of the load-carrying strength of pipe culverts. Most of those on the market have been so widely used, and under such varying conditions, that if properly placed they are likely to withstand the pressure from any loads they may be called on to carry.

The concrete pipe, which is not as generally known as the other types, has many advocates among those who have used this type. Where the material is convenient and available these culverts are said to cost about one-fourth the cost of cast iron. They are made in molds, and in sections 4 to 8 feet in length, according to size. They are similar in some respects to vitrified clay pipe except that they do not have the bell and spigot joint, but are made with square or beveled ends. When placed in position the joints, after being brought to the closest possible contact, are covered with a layer of cement mortar.

In most instances where concrete pipe culverts have been used the county or township has established a central plant, purchased molds, and in rainy weather or at other slack periods put in the time making the concrete pipe for future use.

Vitrified pipes for culverts are made in various sizes ranging from 12 to 36 inches and in sections 2 feet long. In laying these sections the bell end of the pipe should always be up stream, and the best authorities claim that the top of the pipe should be at least 15 inches below the surface of the road. A greater depth is desirable, especially if the road carries heavy loads, such as traction engines or other vehicles with heavy wheel pressure.

Cast iron pipe is usually made in 12-foot lengths. Quite recently, however, cast iron pipes have been put on the market in 4-foot lengths. The standard pipe is uniform in thickness of shell and weight per foot, according to size of the pipe. The new cast iron pipe has a thinner shell, is lighter in weight, and is reinforced with projecting ribs. While the standard pipe has been in use many years, the new pipe may or may not fulfil the requirements. Standard cast iron pipe is very strong and may be placed within 6 inches of the road surface without danger of damage under any ordinary conditions.

Corrugated metal pipe culverts are made in a variety of sizes and of different metals, principally wrought iron and steel. The wrought iron is held to be the more durable, being less likely to damage by corrosion. Its weight is about one-twentieth that of cast iron, which fact applied to freight rates determines its availability in many cases. At the point of manufacture it costs about the same as vitrified pipe, the difference between them in price in any locality depending mostly on the cost of transportation. The vitrified pipe weighs several times as much as the corrugated metal.

Both the up-stream and down-stream openings of pipe culverts should be protected by face walls of concrete or of stone or brick masonry laid in concrete mortar. The face wall at the up-stream or entrance end should be so shaped and placed as to lead the water naturally into the pipe, and so as to prevent the possibility of the water getting around or under the pipe to cause a washout. The face wall at the outlet or down-stream end should protect the embankment from wash and consequent caving, and should also prevent seep-age back under or around the pipe, which would be sub-ject to the destructive action of frost. Neglect of this precaution has ruined many sections of roadway and made expensive repairs necessary.

In laying pipe culverts it is important that the bed be properly prepared. The beds at joints are es-pecially important, as any settlement will throw the pipe out of a true line, and probably cause a leak which is likely to undermine the culvert and the roadway. If concrete is being used on the job, or is convenient, it is well to make a concrete bed for the joints. Otherwise a flat stone, thoroughly imbedded, or earth so thoroughly rammed down as to be prac-tically solid may answer the purpose. After placing the culvert the earth should be thoroughly tamped under and around the pipe. For this purpose sandy gravel is preferable if it can be procured readily.

The slope at which a pipe culvert should be laid de-pends much on local conditions. It should always be steep enough, however, so that the culvert will be self-cleaning and that no sediment shall remain in it. Cul-verts which are smooth on the inside require less slope than corrugated metal, the corrugations having a tend-ency to retard the flow and thus permit the deposit of sediment. Generally 1 inch in 5 feet is as slight a slope as should be provided, while twice that, or 1 inch of decline to 30 inches of distance, will be better.

Box culverts are made of two side walls and a covering over the top. If the side walls be of stone they should be of sufficient thickness so that the earth pressure from back of them will not force them out of position. Where the walls are of concrete they are usually made 4 to 8 inches thick, depending on the height, and are set into the ground a distance of 18 to 24 inches below the bed of the stream with a footing of about twice the width at the bottom. Only in large culverts is it considered necessary to put reinforcing material in the side walls.

If the soil be of a character that is easily washed out, a bottom, or floor, should be made. This may be made of flat stones carefully fitted together, or of cobbles set closely, or of concrete. If, however, the bottom of the stream is hard it is often unnecessary to put in any artificial bed. It often occurs that a culvert is built over a small stream in which for years there had been no material change. In such cases there seems no good reason why any expense should be incurred in the construction of a floor. When a concrete floor is necessary, concrete side walls need only be carried to the bottom of the floor.

The top of a box culvert may be of stone slabs, or

concrete slabs placed on iron I beams, or of reinforced concrete slabs. In case of large culverts the reinforced concrete slabs are usually supported by I beams. These slabs are made 6 inches thick and the reinforcement may be of any one of several approved types. The concrete mostly used is either a $1:2:4$ or $1:2\frac{1}{2}:5$ mixture, and the slabs should season well before being subjected to heavy loads.

The wooden box culvert should not be built except in localities where it is the only available material. It is the most expensive of all culverts to maintain, and its tendency to get out of order on account of loose planks and other defects produces a definite and grave danger to traffic. Its use can at any time only be justified as a temporary expedient.

Arch culverts, whether of stone, brick, or concrete, are built over an arch form of whatever shape of opening may be desired. Unless the bed of the stream be of exceptionally firm material it is usual to put in a floor the width of which is sufficiently great so that it will carry the side walls. Stone and brick are laid in cement mortar and the masonry properly keyed.

In building concrete arch culverts most highway officials provide themselves with collapsible iron forms, which greatly reduces the expense when there is any considerable number of culverts of a similar size to be constructed. There are several of these collapsible forms on the market. They are set in proper position either on the stream bed or the culvert floor. The side forms for the concrete are built of wood in the usual manner and the concrete placed and tamped into position until the arch form is covered to the required depth. As soon as the concrete has set, the arch form is collapsed and withdrawn and is ready for use elsewhere.

In this type of culvert some road officials use reinforcement and others do not. The question seems to be as to whether it is more expensive to put in reinforcement and make the walls and arch thinner, or use more concrete and do without the iron. It is probable that in regions where excessive frost is encountered it may be safer to use reinforcement. In other cases it does not seem necessary. A design of arch culvert much used in Massachusetts is without reinforcement and has given universal satisfaction.

All culverts should have adequate face walls and wing walls to protect the embankment and to prevent water from getting behind the culvert walls. The face walls should be carried down at least 2 feet below the bed of the stream, and lower than that if the soil be of an especially permeable character. If the culvert be the outlet of a side ditch, the outer wing wall and the floor of the culvert should be extended so as to reach beyond and under the ditch. These should be a continuation of the walls and floor of the culvert itself, so that the heaviest flow may not cause damage. At the outlet end the face and wing walls should be so constructed as to protect the embankment and assure the proper run-off for the water.

Bridges are of such variety in type and must meet such varied requirements that only the general principles of their construction may be considered within the limits of this chapter. The materials are wood, steel, concrete, either plain or reinforced, and stone arches.

The stone arch bridge has the sanction of history. There are stone arch bridges now in existence and carrying the traffic placed on them the history of which runs so far into antiquity as to be lost in its oblivion. For some unexplained reason modern engineers do not seem to favor stone arches. That they are practically indestructible, and that they are economical in those sections of country where stone is available, must be admitted.

Wooden bridges have a temporary value only. While the first cost is less than that of those built of more durable materials, the cost of upkeep is excessive and under modern traffic conditions their life is likely to be short. In some instances, however, it occurs that timber is plenty and cheap, and that other materials are expensive and difficult to get, so that a wooden bridge may well fill the gap and answer necessary purposes until a more permanent structure can be constructed. For short spans wooden bridges are usually built with joists extending from pier to pier or abutment, with a plank floor spiked to the joists. For longer spans there is a great variety of trusses to be selected from to secure that which may best fit the local condition.

Iron and steel bridges are constructed in almost an infinity of styles and types. Until within a few years most of the important structures erected during the past half-century have been of iron or steel; iron in the earlier years, steel in the later. For a considerable period of time, from about 1880 to about 1900, a condition prevailed which gave rise to numerous scandals in various parts of the country by reason of means used by steel bridge salesmen to secure several times the value of the bridge by imposing on either the ignorance or the cupidity of local road officials. It was not unusual for a bridge which would cost less than $1000 to be sold for $4000, $5000, or even $6000, the representatives of the various companies deciding who should make the lowest bid, and each sharing in the profits. The system has been practically broken up.

Nearly every bridge manufacturing company makes a number of standard styles and types of bridge which they can supply at reasonably short notice in different length of span at reasonable prices. Modification of these standard bridges may be secured at a small additional expense.

The plans of these standard bridges usually denote the guaranteed load-carrying capacity. This is an important factor, as no highway bridge should be erected without a capacity for carrying a concentrated live load of at least 20 tons with a reasonable margin of safety beyond that weight.

The custom has prevailed extensively, and still prevails to some extent, of accepting the advice of the contracting company's representatives as to the best type of bridge for a given locality; whether girder, pony truss or through truss, steel arch, etc. This is unwise. It must be remembered that the contracting company is actuated by other motives than are the highway officials of the community; and that even though perfectly honest, its recommendation might be influenced by conditions prevailing in its shops or in its business, rather than by the welfare of the community purchasing the bridge.

The life of a modern steel bridge when properly cared for is estimated at from forty to fifty years. For this period of existence, however, the words "properly cared for" must be given their full significance. Thousands of steel bridges have gone to pieces in fifteen or twenty years, and sometimes less, because they were not given proper attention—properly cared for.

Steel when not protected from moisture is subject to rapid deterioration by oxidization—rust. This is especially true in places where the metal comes in contact, but not close enough to keep out the moisture, as at the joints in a bridge. Those parts exposed to the air, with alternate wetting and drying, suffer but little damage; but in the joints the rust is likely to eat into the heart of the metal and, without the damage being visible, produce a condition of dangerous weakness which is hard to detect until an "accident" happens.

The care of a steel bridge includes keeping covered with special paint all joints and connections, so that the moisture shall be entirely excluded. The entire structure should be painted at least once a year; but the joints should be gone over at least twice a year, and oftener if the conditions of heat or cold have suggested breaks in the paint covering. If the bridge has any considerable area of riveting, the rivets should be gone over at least once a year with a hammer by some one who can tell a defective rivet by the sound when the hammer strikes it. By following these principles, and such others as the style of bridge may suggest, the greatest use possible may be secured from the structure.

Concrete bridges have rapidly come into favor in recent years. The facts that there seems no limit to their life if properly constructed; that they can be made more attractive, so as to blend more harmoniously with

(Photo by Chapin, Jacksonville, Fla.)

surroundings; and that as a general rule the first cost is less than that of other comparable types, have given an impetus to their adoption throughout the entire country which may be considered remarkable.

For long spans and large bridges the services of competent engineers are essential. Whether the bridge

Fig. 13.—Illinois "short span" bridge.

be of the beam, girder, or arch type, the very nature of concrete makes necessary a careful expert study of stresses and other conditions connected with the live and dead load that the bridge must carry. And to meet these requirements must be studied the character and quantity of the reinforcement and its proper placing; the particular mixture of concrete necessary to give the proper strength under each individual condition; and a variety of other factors relating to the strength, balance, and durability of the structure. Also to combine these structural elements with a design which shall be graceful and attractive and blend artistically with the surroundings; and a finish which shall be pleasing

Fig. 14.—Concrete pile, cap and slab bridge over Bijou Creek, Colorado.

to the eye. All these factors require engineering intelligence of the first class.

Forty-four of the forty-eight States of the Union have State Highway Departments with competent engineers. In nearly all of these states it is a part of the duties of the department to furnish to counties, townships, or other subdivisions of the state information and advice, and in many states plans of bridges, and to supply expert inspectors to see that the work of contractors is properly performed. Such advice and assistance should be utilized to the fullest possible extent, especially in the case of long-span and large bridges, whether of steel or concrete.

Fig. 15.—Concrete girder bridge, Colorado.

Short span concrete bridges, by reason of their low first cost, their strength, and their presumably lasting qualities with low maintenance charge, have practically revolutionized the practice in short span construction during the ten-year period from 1906 to 1916.

While these are of almost infinite variety and plan, by far the greater number have been built from plans developed by Mr. A. N. Johnson, formerly State Highway Engineer of Illinois. In that state alone more than a thousand of these bridges have been built, and in many other states they have been extensively adopted.

Fig. 16.—Type of "sectional" bridge.

This plan provides for a combined reinforced concrete girder and floor. Only two girders are employed, and these are above the floor, forming the side-rails of the bridge. For this reason the width of the roadway of the bridge is limited to about 18 or 20 feet. The extreme length should not exceed 40 or 50 feet, and the plan is most useful for spans of 30 feet and under.

The amount of reinforcement is adjusted to the length and width of the bridge at a fixed scale. The steel in girders is so arranged as to carry the ultimate weight of the bridge and load, and also to support the transverse reinforcement of the floor of the bridge. The floor, however, has reinforcing steel extending lengthwise as well as crosswise.

The concrete used in these bridges is of a mixture of 1 part cement, 2½ parts sand, and 4 parts of broken stone or gravel. The stone or gravel must not be less than ⅜ inch nor more than 1 inch in size. The comparatively large amount of sand as compared to the stone is considered necessary because of the large area of surface exposed to the air, and also to insure solidity about the reinforcement. This proportion makes an excess of mortar which adds to the holding power of the concrete on the reinforcing rods, and also makes a denser surface, which adds to the appearance of the structure, and at the same time makes it more impervious to weather conditions.

These bridges are designed to carry a live load of 24 tons distributed on two axles 10 feet apart, with 8 tons on the front and 16 tons on the rear axle. This live load is in addition to the "dead" load, or weight of the bridge structure.

Plans and specifications for these bridges in lengths up to 40 feet may be obtained from the Illinois Highway

Department, Springfield, Ill. Most of the details may be found in the various reports of the department together with costs of construction of various bridges. These latter vary so widely, however, owing to variance in conditions that they hardly furnish a guide.

On most bridges, even on wooden structures, the tendency in recent years has been to use concrete floors. These are easily laid, and especially if reinforcement is used, most experts hold that the added dead weight on the bridge structure is more than counteracted by the rigidity which the concrete floor gives, with the resulting reduction of vibration. Also, a concrete floor lasts many times as long as a plank floor, and the cost of its upkeep is negligible, making it much cheaper in the long run.

It is usually considered that a good concrete foreman is capable of building the short-span concrete bridges above described, and of placing concrete floors on existing structures without further engineering directions than are found in the plans and specifications. Inspection by a state highway official, if available, is, however, always wise.

Foundations for highway bridges have in recent years followed closely the principles for railway bridges. It has been found in many instances that a foundation that would carry the weight of a bridge and load might be easily undermined by flood water in case of a freshet.

It is, therefore, considered the best practice, if bed-rock cannot readily be reached, to drive long piles to a practically solid bearing, and place the masonry or concrete abutments and piers on the top of the piles. Wooden piles should be cut off some distance below the low water level, and generally below the bed of the stream. Concrete piles are often used, formed by driving down a steel core with a sheet iron shell, then withdrawing the core and filling the shell with concrete. In this form of pile it is possible to have the bottom of the shell opened and a quantity of concrete forced out into the surrounding earth if it is soft enough, making a broader foundation, the pile taking the shape of an inverted mushroom. Other styles of piles are made, with or without reinforcement, and either driven to place by specially prepared pile-drivers, or sunk into place—in sandy earth—by the action of a small jet of water at high pressure at the point of the pile.

Steel caissons with a timber framework at the bottom and concrete at the top have been much used. These are sunk into mud bottoms as far as possible, and the piers and abutments built on them when they have reached an apparently secure resting place. There is often danger that the resting place is not secure. Caissons are sometimes sunk to bed rock or to a solid foundation, the space within the caisson excavated during the sinking operation, and then the caisson filled with stone, concrete, or such other material as will make an absolutely solid base. The individual treatment in each case depends on the soil and other conditions encountered.

Piers and abutments should always be parallel with the flow of the water. If the bridge must cross the stream diagonally, the skew must be in the bridge rather than in the piers and abutments. Girder bridges built on a skew are readily adjusted; arch bridges on skew present technical features in stress distribution which require accurate engineering study.

The basic factor in securing a satisfactory pier and abutment foundation for a bridge is to get to a depth or to a condition of solid earth or rock, so that they will neither settle nor wash out.

Floorings on bridges may be made of any desired material, usually governed by the cost. Concrete bridges and many steel bridges have concrete floors. Sheet asphalt, asphaltic concrete, brick, wood block (creosoted), and a variety of other floors are in general use. A study of which is best adapted to the traffic and, therefore, most economical, under local conditions, is necessary for the determination as to which is most available.

CHAPTER VIII

ROAD TRAFFIC

Before beginning the construction of a new road or the reconstruction of an old one a careful study should be made as to the amount and character of the traffic which the road will have to carry after its improvement. The importance of this study cannot be overemphasized, as there is always a great increase in traffic after a road has been improved. Travel will invariably go a considerable distance from its most direct route if by so doing it can reach a good road.

Previous to the general use of motor cars the changes in courses of travel could be readily estimated. Since the tremendous development of this method of transportation the subject is one which calls for the best and most careful consideration by road officials. The road which is sufficiently well built so that it will carry light and heavy horse-drawn traffic may go to pieces directly under the motor truck or the motor omnibus. It is worthy of note that with the improvement of the roads motor-bus and motor-freight lines are springing up by hundreds, furnishing regular passenger and freight schedules to communities either off the line of a railroad, or where the local railway service is unsatisfactory.

The traffic census, that is, counting the vehicles and animals passing over the roads, was first practised in France about 1840, though the custom had prevailed in separated localities previous to that time. Since then there has been a regular census of traffic taken over all the roads in that country at intervals ranging from six to ten years. Since the advent of the automobile the traffic on certain roads where the travel is heaviest has been taken at more frequent intervals in order to ascertain a basis for determining the proper kind of repairs necessary to preserve the roads.

In England for many years a traffic census has been carried on by County Councils on various roads when there seemed to be a need for it. In 1912 control of the traffic census was taken over by the Road Board, and is now conducted regularly and simultaneously throughout that country. The actual work of making the count is still in the hands of the County Councils, but under the direction of the Road Board, thus securing uniformity of method and dates of counting, and of classification of the units or items of traffic.

In this country only occasional counts were made of road traffic in various localities until 1909. During that year the Massachusetts Highway Department organized a traffic census covering a large portion of the main roads of that state, and has kept it up at three-year intervals. The same year the Illinois Highway Department caused a census to be taken of the traffic at a large number of important points, at many of which the traffic has been again counted at such intervals as the conditions seemed to warrant.

As an illustration of the remarkable change which sometimes occurs in the volume of traffic over a road, the chairman of the Massachusetts Highway Commission, in an address at the Pan-American Road Congress (Oakland, California, 1915), mentioned one of the roads of his state. The road mentioned was 26 miles long. In 1909 the traffic census showed it carrying 37 automobiles a day, as well as a fair amount of horse-drawn traffic. In 1912 the number of automobiles daily was 250, and in 1915 the count showed a little more than 1000 automobiles a day, the horse traffic remaining about the same or a trifle less than it was in 1909. The especial point of the remark was that the road, having a gravel surface, would carry the traffic economically up to and past the 250 motor cars per day, but when the number reached anywhere near 1000 (probably near 500) the cost of keeping the gravel road in condition was so great that economy demanded that the road be rebuilt with a stronger and more durable surface.

It is, of course, impossible to form an accurate estimate at all times. Sometimes new conditions will arise by the construction of bridges in new locations, or by the organization of through routes of travel, or by the development of new industries made possible by the improvement of the roads. But a fair judgment can be made in most instances by taking a census of the present traffic and using it as a basis from which to figure the probable accumulations from parallel routes and from additional accessible communities after the road shall have been improved.

In taking a census of the traffic two methods are in use, each having its advocates: One method is to set apart a week in the spring and a week in the fall, and have the count made during seven full days at each period, the hours being usually from 7 A. M. to 7 P. M., though the hours may be changed to suit the particular travel habits in various localities. The other method is to use seven weeks in making the count, taking Sunday of one week, Monday of the next week, Tuesday of the next, and so on until a complete week has been covered. Advocates of this latter method claim that it gives a fairer average by distributing the count over a greater period of time, while those favoring the first method consider that it is possible to secure a better organization for the work if it is done on consecutive days.

In taking the traffic census arrangements are usually

made with some resident of the immediate vicinity of each point decided on. In many instances, especially in those months when schools are not in session, students are employed for the purpose. Many persons, recognizing the public nature of the work of making the count, volunteer their services gratuitously; others will accept but a small compensation. In any case it is necessary that the sense of responsibility of the person making the count be sufficient to insure thorough and reliable work and results.

The points fixed upon for counting the traffic are usually a little out of the city or village—far enough to avoid the traffic which belongs in the city and yet near enough to include all the outside traffic to or from the city over that particular road. When the points have been selected and the census takers duly appointed, each census taker should be supplied with printed instructions and blanks which should be prepared in such simple form as to make mistakes practically impossible.

Different states and different countries have each their special classifications of vehicles. Some are very elaborate, with fifteen or twenty separate items. For all practical purposes, however, the following division of the traffic seems sufficient:

Automobiles:
Runabouts,
Touring cars,
Motor trucks.

Horse-drawn vehicles:
- 1-horse light,
- 1-horse heavy,
- 2 or more horses light,
- 2 or more horses heavy.

Special:
- Traction engines,
- Motor omnibus,
- Freight motor with trailers.

With the list printed on the left of the sheet and wide lines left for marking a "/" in the proper place at the passage of each vehicle the work is easily performed. On many country roads the counting has been done by housewives without materially interfering with their household duties. It was only necessary that they keep in sight of the road. At the close of each day and at the end of the period the traffic can be footed up and the proper comparisons made.

As to what kind of a road to build and how to build it, after the actual traffic has been counted and the future traffic estimated as carefully as possible, there has been much experimenting, and many hasty conclusions formed. There are some cases, however, where experiments have been carried on over a series of years and the results carefully recorded and made use of. Of these, the reports of the Massachusetts Highway Department furnish the most comprehensive and reliable record reported, both on account of the continuance of the experiments through a series of years and of the practical use made of them under the constantly changing conditions. It is possible that other equally valuable information may have been secured, but not reported in form to make it available to the public generally.

The Massachusetts Department says in its report:

> A good gravel road will wear reasonably well and be economical with a traffic of 50 to 75 light teams (1 or 2 horses), 25 to 30 loaded 1-horse teams, 10 to 12 loaded 2-horse teams, and 100 to 150 automobiles. If the number of automobiles averages above 150 per day the road should be oiled.
>
> An oiled gravel road, with good oil applied either hot or cold, ½ gallon to the square yard (cold oil must be used yearly), will carry a traffic of 75 to 100 light teams, 30 to 50 loaded 1-horse teams, 20 loaded 2-horse teams, and 500 to 700 automobiles of all classes.
>
> A water-bound macadam road will stand a traffic of 175 to 200 light teams, 175 to 200 heavy 1-horse teams, 60 to 80 and perhaps more heavy 2-horse teams, but not to exceed 75 automobiles, especially if they move at fairly high speed. A dust layer will be serviceable when the automobiles range from 50 to 100. With a good dust layer kept well applied the road will stand 300 to 500 automobiles a day in addition to the team traffic, although the stone in the macadam will wear.
>
> Water-bound macadam with a hot oil blanket coat will be economical with 150 to 200 light teams (1 or 2 horses), 75 to 100 loaded 1-horse teams, 25 to 30 loaded 2-horse teams, and automobiles up to 1400 in number. The large number of pneumatic tires has been found to iron out the depressions caused by steel tires in the road and keep it rolled down smooth. The road will not be injured by 50 or perhaps more motor trucks, but it will crumble and perhaps fail with more than 100 loaded 1-horse and 50 or more loaded 2-horse teams, such as loaded farm wagons, ice wagons, wood wagons, etc., especially on narrow tires.

(All the figures given in the statement represent average daily traffic, and in most cases seem to be based on conclusions drawn from examples of maintenance of old rather than of newly built roads. This seems especially probable in view of the fact that the Massachusetts Highway Department no longer builds water-bound macadam surfaces on its main roads, but employs more durable surfaces exclusively.)

In some states laws have been enacted limiting the loads to be carried to 800 pounds per inch of width of tire. Narrow steel tires on heavily loaded wagons or trucks are considered the most destructive agency to which the roads are subject. Solid rubber tires on motor trucks are usually constructed of much greater width than would be required by such a law, and the rubber does not have the same destructive force as steel tires; while a pneumatic tire of whatever size flattens out under a load, and presents an enlarged surface which is beneficial to the road surface rather than otherwise. But steel tires under heavy loads and with sharp corners cut into the road surface, leaving depressions where water accumulates; then more traffic churns the water, and soon there is a mud-hole, and a rapid softening and displacement of the material forming the road surface.

In some instances computations have been made to determine the weight of traffic per foot of the width of the road. This practice is necessary for a complete knowledge of the stresses on city streets, and perhaps on a few broad boulevards where the entire road surface is in constant use; but on country roads, where the vehicles invariably follow each other in a single or a double track, the knowledge gained will be useless and confusing.

In taking a traffic census special attention should be given to the extra heavy vehicles, such as traction engines, road rollers, etc., and some of the very heavy motor freight trucks which are rapidly coming into use, Usually these vehicles do not damage the road surface on account of the width of the tires, having a tendency to compact and improve the surface; but except in the case of very well built roads their weight is liable to fracture the foundations. There are many of the older bridges on the roads throughout the country which will not stand under these heavy weights. Until about ten years ago most highway bridges were built to sustain not more than 4 to 8 tons. Such bridges, until replaced by stronger ones built with reference to the weight of modern vehicles, must either be braced or in some way given additional support; or else the heavy vehicles must be kept off them. Slats or other projections should not be permitted on traction engine wheels on any hard surfaced road.

Traffic regulations are usually fixed by law, the laws

varying in the different states. Speed limitations, unless within liberal limits and liberally construed, have generally been found vexatious and unprofitable. The most satisfactory method of controlling the speed question has been found to consist in demanding and enforcing a *reasonable* speed, according to conditions.

A speed which might be reasonable at one point on a road might be entirely unreasonable at another point; or which would be reasonable at one hour of the day, unreasonable at another hour. On a wide straightaway road, with no road-crossings, and no obscurities for other traffic, and with pedestrians absent, almost any speed at which a motorist might wish to travel could be considered reasonable. If there were frequent cross-roads, or much travel, or any other factor which would make speed dangerous to people or property on the road, then the speed should be reduced until all danger should be eliminated, even to the speed of a man walking. Responsible motorists are the first to condemn reckless driving, and punishment of offenders should be based on recklessness in driving rather than on miles per hour.

In states where the licensing of automobile drivers is in the hands of the State Highway Department the question of safe driving is much more easily controlled than where the roads and licenses are under separate control. Reports of violations, accidents, etc., all reported to the common department head, gives a more comprehensive knowledge of those individuals who habitually violate the rights of others, and permits licenses to be cancelled when necessary without very much "red tape."

One broad principle in connection with road traffic must always be borne in mind: No single individual has the right, legally or morally, to either damage the road, which belongs to all, or to so use it as to endanger the life, property, or happiness of another.

CHAPTER IX

ROAD FINANCE

There are two basic factors in road finance: getting the money, and spending it to the best advantage.

Getting the money involves the questions of valuation, taxation, bonds, either sinking-fund or serial, and outside aid or contributions. Spending the money takes into consideration the advantages or disadvantages of location regarding materials; freight rates; the most economical kind of road for the traffic; and the local laws respecting bids and bidding.

Getting the Money

In some well-settled and wealthy communities where land values are high direct taxation for road building is considered wise. But the localities where such taxes can be levied without being burdensome comprises a very small percentage of the country. Almost everywhere, if good roads are to be built, the cost must be distributed over a series of years. The length of time over which the cost must be spread depends on the sentiment of the community and the local laws regulating bond issues. In issuing bonds consideration must be given to the fact that the people have the use of the improved road while paying for it.

In some states there is a provision for five- and ten-year certificates which those directly involved may take advantage of. This, however, is usually confined to the few localities where the cost of construction, or some part of it, is taxed against the adjacent property. Taxing the adjacent property to build roads for all the people is recognized in modern times as a wrong principle, and has been discontinued in most of the states.

Generally, bonds are of three classes: Sinking fund bonds, serial bonds, and annuity bonds.

Sinking-fund bonds are those which run for a stated term of years, and which require that a certain percentage or a certain amount be set aside every year as a sinking fund for the payment of the principal at the maturity of the bonds. There is often misapprehension or faulty figuring on the amount to be thus set aside. The sinking fund should be so invested from time to time as to draw compound interest at the best rate that can be obtained. In many counties the county boards invest the money of the sinking fund in other local securities, thus getting the same rate of interest on the sinking fund money as is being paid on the bonds. One county which came under the author's observation was able to take a full issue of $100,000 of road bonds with the money in the sinking funds of other county securities.

These funds had been drawing 3½ or 4 per cent. interest. By investing in the issue of road bonds they drew 6 per cent., and only the interest was left to be invested annually or semi-annually until the maturity of the bonds.

Compound interest tables will show what amount invested annually at compound interest will reach a given amount in a given period. For instance, 1 per cent. of the face value of a bond invested and compounded annually at 3½ per cent. interest will produce enough money to pay the bond in a little less than forty-eight years. This amount is frequently used, and collected at the same time as the interest, when fifty-year bonds are issued. With a higher rate of interest, such as would be obtained by investing in the same or other securities, the term would be shortened or the annual sinking fund investment reduced.

Compound interest tables, which are easily procured, will show the amount necessary to be deposited in a sinking fund at any ordinary rate of interest for any general term of years. Nearly all banks have these tables, and they are in the rate book of nearly every life insurance company. They can be purchased for a small amount at almost any news-stand. The proper table is easily applied to any condition of term of bonds and rate of interest, so that every one interested may easily secure complete information on the subject. The tables are too long for reproduction here.

Serial bonds are in much favor in some localities. These bonds are numbered, and are usually made in uniform amounts. An issue of $100,000 might be made in 100 bonds of $1000 each or, if intended for subscription by the people of the community, might be made in denominations of $100 each, which would cause the numbers to run to 1000. This applies to all cases where it is desired to give an opportunity to local investors, both of large and small means, to subscribe to the bonds.

With serial bonds, in addition to provision for paying the interest, the money for the principal instead of going into a sinking fund is used to pay the bonds according to their serial numbers. As an instance, $100,000 of bonds might be made payable at the rate of $4000 each year for twenty-five years, or $5000 each year for twenty years; or a period may be chosen which, by paying $20,000 or $25,000 at each time of payment, will reach the final payment in twenty-five or twenty years.

Take $100,000, to be finally paid in twenty-five years. The annual interest at 5 per cent. will amount during the first five years to $5000 per year. At the end of the first five years the first twenty of the $1000 bonds are to be paid and provision must be made for $20,000 required for this payment. During the next five years the interest is reduced to $4000 a year, or 5 per cent. on $80,000. At ten years the next twenty bonds are paid, reducing the interest to $3000 a year, and so on, until after the twentieth year the interest amounts to but $1000 a year, and at twenty-five years ceases entirely, when the last twenty bonds are paid, finally cancelling the debt.

In former years trust companies and large bond-purchasing concerns did not look with favor on serial bonds, preferring the long-time sinking-fund bonds which gave them less trouble in re-investing the funds under their control. In some instances there may have been a fear of reducing the amount of sinking-fund money which might be placed in their hands on deposit at low interest rates. In recent years, however, these financial institutions have taken more kindly to the serial bonds, so that at the present time (1917) they command a price in the larger financial markets fully equal to those of uniform term.

Annuity bonds, while of long and favorable use in private business and real estate transactions, are comparatively new in their application to public securities. The plan is to so figure the interest, principal, and term into a system that all the annual or semi-annual payments shall be practically the same, so that the debt will be paid in a given number of equal installments. To illustrate the working out of the annuity bond system the accompanying table has been rearranged from a scientific document published by the United States Department of Agriculture. The table is based on a loan of $100,000 payable in three years, in six semi-annual payments, the interest being at the rate of 5 per cent. A simple computation in partial payments will adapt the plan to any other amount for any period of time and at any desired rate of interest. The table follows:

Time.	Principal outstanding.	Interest at each payment.	Principal paid at each payment.	Semi-annual payment.
6 months	$100,000.00	$2500.00	$15,655.00	$18,155.00
1 year	84,345.00	2108.63	16,046.37	18,155.00
1½ years	68,298.63	1707.47	16,447.53	18,155.00
2 years	51,851.10	1296.28	16,858.72	18,155.00
2½ years	34,992.38	874.81	17,280.19	18,155.00
3 years	17,712.19	442.81	17,712.19	18,155.00
Totals		$8930.00	$100,000.00	$108,930.00

Examination of this table shows that the reduction of the principal amount increases as the interest payments become less, the first payment on the principal being $15,655, while the last one is $17,712.19.

Some communities seem to favor the annuity bond plan, as it gives them an exact idea of what they must pay each year, and because it is the same amount at each payment period; while in serial bonds the interest payments are heavier at the beginning, and become smaller as the bonds are paid off; and with sinking-fund bonds much depends on the rate of interest that can be secured for the sinking-fund money.

Misapprehension of the expense of bonds and the amount of value to be derived from their issue exists in some localities where bonds for roads have been defeated. A simple form for determining the cost to each taxpayer is to get as accurately as possible the number of acres which will be taxed to pay the principal and interest on the bonds annually, and figure what will be the average cost per acre per year. The result will surprise many who object to bonds. Another plan is to take the amount which must be paid annually on the bonds, principal and interest, and divide it among the total assessed valuation of the district involved so as to get the rate of additional taxation. This rate extended in any individual case will show the amount of extra taxation which the bonds will involve. In most cases the amount will be so small as to silence objection to the bond issue.

On some roads in some sections of the country private individuals, corporations, automobile clubs, and other corporations are found willing to contribute certain amounts toward the cost of a road, either as a matter of creditable public spirit, or for furthering an improvement which may benefit them, or for both reasons. The roads built with the aid of such donations are constantly increasing in number, and indicate a growing intelligence on the subject of the value of improved roads to a community.

Statistical experts have figured that the building of a good road leading to a market town or business center increases the value of the farm land accommodated by the road from 15 to 50 per cent., and cases are not rare where land has even doubled in value within a year after a road was built. This increase is due principally to the decreased cost of hauling crops and supplies over the road, and the larger number of crops that can be profitably grown by reason of cheap hauling at all seasons of the year.

Spending the Money

The type of road to be economically built in any community should be determined as far as possible by the availability of materials. In almost every section there is some material which, either by itself or in combination with some other material which is not too expensive, will make a fair road at a reasonably low first cost. If the location be one where a high-class road is necessary and there is no suitable local material, the added cost of shipping in outside materials must be considered as part of the road investment.

The points on which to figure are: The relation of the first cost and the cost of upkeep for a series of years of the road built cheaply and of inferior materials, as compared with the first cost and the cost of upkeep of the more expensive road.

A good road, made of first-class materials, should require but a small outlay for repairs for a number of years; while the road built of inferior materials will require repairs more or less constantly, according to the quality of the material and the character of the traffic. In this connection it is held that when the annual cost of keeping an inexpensive road in repair amounts to enough to pay the interest on the money necessary to build a good road, it is economy to build the good one. The same rule will apply to all classes of roads, wherever located. The particular advantage is that without additional cost a good road may be had instead of one which is in a condition where constant repair is necessary.

The laws of various states vary greatly as regards construction. In many states all construction must be let by contract; in some states the lowest bid must be accepted; in some states officials can build the roads on force account, and in some convicts may be employed.

Where construction is done by contract the specifications should be carefully drawn before bids are invited, so that no materials or workmanship may be employed of a lower grade than that intended. In some materials which have names which indicate their general character there is a vast difference in excellence. The word "asphalt," for instance, covers all materials with an asphaltic base, whether mined from the asphalt lakes of South America, or those which are the residuum of petroleums from the oil wells of the United States and of Mexico. Care should be taken, after investigation, to specify as accurately as possible the exact material required whenever it is possible to do so.

There are three definite kinds of specifications, either of which may be employed where the law permits: One is the "open" specification, sometimes called a blanket specification, the limits of which are so broad as to admit all classes of materials and types of road. This gives an opportunity for competition in materials as well as price, and gives the road authorities an opportunity to study the respective advantages of the different materials and roads at the prices named. But this "open" specification may not be used when the law requires the acceptance of the lowest bid, as that would necessarily mean the poorest road, of the cheapest material. If the road officials have undisputed authority in the matter of accepting what seems to them the best proposition, the "open" specification has many advantages.

The "closed" specification enables the road officials to get exactly the materials they want, as the specifications are drawn with reference to that particular material. Objectors to the closed specification claim that this form limits competition and tends to higher prices, by reason of keeping out other competitive materials. While this may be a fact in isolated instances, it is a fact that nearly all large manufacturers of road materials will sell to all contractors at the same price; so that competition must be between the contractors for the work and other factors entering into the construction. As before stated, in the opinion of the author, especially as to such materials as brick or asphalt, the specifications should state definitely the particular material desired.

"Alternate" specifications, which describe two or more types of roads or classes of material, are sometimes used. This "alternate" specification has the effect of enabling the officials to receive bids by which they can balance against each other the relative advantages and price of the different types of road. It may also be drawn so as to describe different classes or grades of the same material, and thus the officials may have a chance to decide what particular material at the price bid will be of the greatest advantage to the community. The "alternate" specification has the effect of a series of "closed" specifications competing against each other.

Where work on road construction is done under the supervision of local officials on force account the most careful records of all costs should be kept and properly classified, not only that the public may be kept fully posted, but that other or subsequent officials may have the benefit of the information they contain.

Nearly all the states of the Union grant state aid in road building in some form, either cash, or convict labor, or engineering services, and sometimes in two or more of these forms. Such aid should be taken advantage of to the fullest extent, and all local officials and others interested should keep in close touch with the Highway Department of their state.

The Federal Aid Good Roads Act, which became a law in July, 1916, appropriates $5,000,000 for the construction of roads during the year ending June 30, 1917, $10,000,000 for 1918, $15,000,000 for 1919, $20,000,000 for 1920, and $25,000,000 for year ending June 30, 1921. It also appropriates $1,000,000 per year for ten years for constructing roads in national parks.

The money thus appropriated is divided among the states in three ways: One-third according to the ratio of population; one-third according to the ratio of area, and one-third according to the ratio of mileage of rural mail routes and post-roads. In establishing the ratios, the population, area, and mileage of post-roads are each figured in relation to the totals of those factors in the United States.

The funds thus appropriated are paid by the Federal Treasury to any state as payment of one-half the cost of construction of any road agreed upon by the Highway Department of the state and the Secretary of Agriculture of the United States, and after the state has provided for the other half of the cost. The roads built with the aid of Federal funds must be of substantial character, and be inspected and approved by Federal engineers. The states must agree to maintain the roads so built.

The office of Public Roads and Rural Engineering, Washington, D. C., has the rank of a bureau in the Department of Agriculture. It furnishes the services of road engineers on the request of local highway officials, and publishes many bulletins and other documents on road subjects which can be procured by those who are interested. Most of these documents are sent free on request; others are subject to a slight charge.

PART II

CHAPTER X

EARTH ROADS

EARTH roads we have always with us, and are likely to have in the future, indefinitely. Of the approximately 2,200,000 miles of roads in the United States, about 2,000,000 miles are earth roads. Some of these have been improved by draining, shaping, and grading, but by far the greatest part of the mileage still consists of the practically unimproved earth road.

The reason for this condition lies in the fact that from 80 to 85 per cent. of the road traffic of the country is carried on from 15 to 20 per cent. of the roads. The main roads, on which the travel comes together from every direction in its movement toward the city, or village, or shipping center, requires a better surface than any of those which lead to them. These main roads are dealt with in other chapters.

The system by which the relative travel over various roads has been accurately determined was worked out some years ago by Mr. A. N. Johnson, then State Highway Engineer of Illinois. He took as a basis an average township, 6 miles square, with a village in its center. At the village are the railroad station, the post office, the stores where hardware, dry goods, groceries, and other supplies are purchased, repair shops of various kinds, the churches, the newspaper, the graded school with high school department, and in some states the township school, which has taken the place of the former district school. The village is also usually the center of social activity for the surrounding country.

With a road on each section line, there would be 36 miles of north and south roads, and 36 miles of east and west roads, 72 miles in all. The center road of the township in each direction, crossing at the village in the center, should be termed the main roads. These would be 6 miles long each or 12 miles in all. This amounts to $16\frac{2}{3}$ per cent. of all the roads, which must be classed as main roads, and improved accordingly. The other $83\frac{1}{3}$ per cent. of the roads, which carry the travel to the main roads, does not require the expensive treatment which must be given the main roads.

Taking each quarter section as a farm, and each farmer making two trips to town each week, it is easily shown that every person in the township uses the main road for at least a part of his trip. The most remote resident travels $2\frac{1}{2}$ miles over the side road and 3 miles over the main road in reaching the village. Forty-four reside on the main road; thirty-six are a mile from the main road; twenty are 1½ miles away, and twelve are 2 miles, making, with the four that must travel 2½ miles to reach the main highway, the 144 quarter sections into which the township is divided. Each of these residents takes the shortest course to the main road, thus aggregating upon it fully 85 per cent. of all the travel.

While each citizen is entitled to a road which will carry his traffic, the expenditure of the same amount of money, for similar improvement, on the road that is used by ten loaded teams per day and on that which carries one hundred such loads would be wasteful. Either more money would be spent on the side roads than is necessary or the improvement to the main road would be of such poor quality that the road would soon go to pieces.

The figures, as given above, are more readily worked out with the township under what is known as the Congressional Survey. In the original thirteen states of the Union, where most of the surveys were more primitive and irregular, careful observation has shown that the relative percentages are practically the same.

The width of the earth road must depend on the present and prospective travel, but should range from 20 to 30 feet between centers of ditches. The right of way usually runs from 40 to 66 feet. The surface of the roadway should be curved or rounded into a crown with the highest point in the center of the road where the road is straight; on a curve the slope should be inward from the outside, and on a side hill the highest point in the crown should be two-thirds or three-quarters of the way toward the outside edge.

The height of the crown of the road should be measured from the bottom of the side ditches, and must depend on the character of the earth of which the road is made. It should be sufficient so that rain or other surface water will readily run off into the ditches, but without damage or wash to the road. It must not be too high, or vehicles will keep to the top, and by following in each other's tracks form ruts; and the water, getting into the ruts, and being churned by other vehicles, will soon cause the formation of mud-holes and a gradual disintegration of the road. Nor shall the road be too flat unless the soil be sandy; because a flat road soon becomes a depressed one, and water will stand on it, and water will destroy the usefulness of any road sooner or later.

After the grade of an earth road has been decided on (see Chapter III), the alignment of the traveled portion to be improved becomes important. It is not unusual to find country roads, extending for miles between straight line fences, winding from one side to the other of the right of way, instead of maintaining a straight line in the center. This irregularity should be corrected when the road is improved. The ordinary way of doing this is by establishing a center line of the road midway between the two fences, setting stakes at frequent intervals, and sighting along the stakes to see that they make a perfectly straight line. Then measurements may be taken from the center of the road to the center of the ditch on each side, and other stakes driven beyond the ditch to indicate the slope of the outer side of the ditch. This slope may be 1 foot back to each foot of depth, or $1\frac{1}{2}$ feet or even 2 feet, according to the soil. In a firm soil a 1 : 1 slope is sufficient; while in a soil which will wash easily a 2 : 1 slope is none too much. The principle involved is that all care must be taken to keep the ditch from filling up or becoming clogged.

In improving the earth road much depends on the location. If the grade requires a cut and a fill in close proximity, ordinary farm ploughs and drag scrapers can be used economically. These carry the dirt from the cut to the fill with one movement. If the distance be more than 50 to 100 yards, wheel scrapers are more economical, as they carry two or three times as much earth with approximately the same horsepower and labor. Wheel scrapers are rarely economical, however, in stony or rocky ground.

Where the ground is practically level the ordinary

road grader can be used to excellent advantage. It is better not to run the blade of the grader too deep at any one cutting, as that has a tendency to throw the earth toward the center of the road in lumps or chunks. Rather, the grader should take comparatively thin depths of earth and spread it thinly in courses over the roadway. This method secures a more even distribution and density of the earth placed on the roadway, and makes a smoother and better road when it is packed down either with a road roller or by the travel.

If the ground be low or swampy so that the road has to be built up, an elevating grader will be found economical and serviceable. These machines not only move earth at a reasonable cost, but they deposit it with such regularity of position that a saving is effected in the subsequent work.

Whatever implements may be used in building an earth road, it is essential that no vegetation be left on or in the roadway. All sods, grass, weeds, bushes, leaves, and every other growth should be placed outside the ditches, so that by no possibility may they get to the roadway. Decaying vegetation attracts moisture; and moisture distributed unevenly in or under the surface of a road soon fills the roadway with holes and irregular depressions, which, if not repaired promptly, soon makes a bad road out of what should have been a good one.

In some cases it may be found that the road can be brought to the required shape and the earth evenly distributed by the use of the grading machine. Generally, however, much may be gained by giving it a thorough dragging with a square tooth farm harrow. This dragging, if done thoroughly, brings about a uniformity of density that is necessary in securing a smooth and even surface. Where the top soil is sand or gravel, and the grader brings up earth from the ditches which has an admixture of clay, a disc harrow should be used in order to thoroughly mix the clay and loam and sand or gravel. This mixture when found makes an excellent road, it having many of the elements of the Sand-clay Road which is described in Chapter XII.

After an earth road has been graded and shaped, and the fresh earth on the surface evenly and uniformly distributed, rolling will add much to its excellence and durability. The question of the availability of a steam or gasoline roller, and the expense connected with its use, has much to do with the economy of rolling an earth road, and each community must decide the question for itself. Thousands of townships and other municipalities throughout the United States own road rollers as a part of their municipal highway equipment; in case of a public-owned roller, or where one can be secured on reasonable terms, a thorough rolling of the earth road will result in an ultimate economy. Even

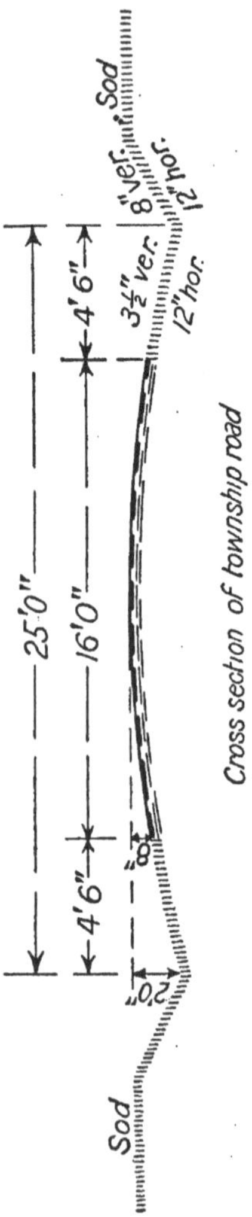

Fig. 17.—This is an approved shape for a road of earth or of a natural gravel soil.

rolling with the ordinary farm roller, drawn by horses, will be a benefit to the road, but not to the extent that a heavy rolling would be. Whenever earth roads are rolled the rolling should begin on the sides, and continue toward the center, so that the top of the rounded roadway will be the last to be rolled.

In improving an earth road, especially where it is to be straightened or other irregularities of subgrade corrected, it is often necessary to use the plough. When hard places are left in the former road, to be covered with different thicknesses of fresh earth, the results will not be satisfactory. The hard surface must be ploughed up to a sufficient depth so that the fresh earth, placed upon it, will wear down into a compact mass. For this reason it is often necessary to plough up the entire road surface, cutting only as deep as may be necessary to get a fairly even foundation, without hard spots or soft spots which may later affect the surface.

The ploughing should be done by back-furrow, throwing the earth together along the stakes recently mentioned on the center line of the road. Care should be taken that each outside furrow reach a trifle lower base level than the one before it, so that the hard earth forming the subgrade may have a gradual slope toward the side ditches. Moisture accumulates more readily in newly placed earth than in the harder undisturbed ground, and if the hard ground is loosened in such a manner as to make a slope from the center each way, the moisture in the fresh earth placed on top will have a chance to drain away. Otherwise, if the hard ground be left higher at the sides than in the central portion of the roadway, the hard ground will act as a basin and keep the moisture confined in spots, which will permit depressions and sometimes holes in the surface. When graders or road machines are used, especial care should be taken to see that there is no comb or ridge of hard earth between the center of the road and the side ditch.

In most kinds of soil, and in the sections of the country where the ground freezes from a few inches to several feet in depth in winter, grass or weeds should not be permitted to grow on the sides of the roadway.

Exception may be made where the road is on an embankment across a low or swampy section where the earth disintegrates or gives way easily. In such cases the formation of a strong sod often holds the earth in place and prevents the road-bed from giving way. Except in such cases all vegetation should be kept off the roadway.

In the Southern States, where the element of frost is not a factor, there are other variations of the above principle to be considered. In some sections of the state of Florida, where the soil consists largely of a very fine sand, and the ground is nearly level, with very slight drainage, the earth road is in its best condition when the moisture permeates its base, and water stands in the side ditches a foot or two below the level of the roadway. The explanation of this seeming paradox lies in the fact that that particular quality of sand packs hard when moist; when it is dry it becomes loose, and wheels plough through it instead of going over it.

This peculiar quality of soil is recognized even by the railroads. Over long stretches of territory in Florida railway embankments are built 3 or 4 feet above the general land level, and the drainage ditches along the sides, while duly equipped to carry off surplus storm water, are so arranged that water will stand in the side ditches and keep the base of the embankment saturated. The particular quality of sand in the locali-

Fig. 18.—Well shaped, graded, and drained earth road improved at a cost of $55 per mile.

ties where this practice prevails is not generally found in other sections of the country.

When an earth road is built on an embankment or fill, the side slopes should be sodded or sown with grass seed so that a sod will form. This will in a large measure prevent the slopes from washing in the run-off of storm water, and from being damaged by wind and other storms. The grass should not be permitted to grow up over the edge of the slope into the roadway, as that would retard the run-off and cause the water to soak into the road. The same rule of growing sod applies when the fill is only on one side of the road; and it is also a good plan to follow for the outside slope of the side ditches, when the road is built in a cut or on level ground, and for the up-hill slope on a side hill road.

The proper time of year to improve an earth road is the late spring and early summer. The work should not be begun on the roadway proper, except where the grade or location is changed materially, until after the ground has settled—after the frost has gone out; and it should not be delayed so that the moisture shall have been dried out by the heat of summer. When the fresh earth is put on the road, harrowed, pulverized, and shaped into the proper crown, it should be moist enough so that either rolling or the travel will pack it down into a hard, solid mass.

If the improvement of the road is left until the ground is dried out, it will be largely a waste of time and money. The travel over it will cause a good deal of dust; will wear the surface down into furrow-like dry ruts, and when the fall rains come the fresh earth put on will be in a condition to absorb and hold the water until it is thoroughly saturated and becomes a mass of mud. If by chance it should freeze while in that condition, the going out of the frost in the spring will leave the road as badly in need of repair as though nothing had been done to improve it. Water has a remarkable affinity with freshly moved earth.

If, however, the road has been improved early in the season, so that the travel of the summer and fall has packed the earth down into a solid mass; and if the maintenance has been kept up intelligently, the fall rains will run into the side ditches, and the frost will have no appreciable effect on the road surface. Then, when the spring comes again, just a small amount of maintenance attention will put the road in first-class condition. With even reasonable care an improved earth road should be at its best the season following its improvement.

While the subjects of grades and drainage are fully treated in Chapters III and IV, there are some features which apply especially to earth roads which may be described in this connection.

Within reasonable limits, as the grade of the earth road increases the crown of the road should flatten. If the crown on a level road be 1 inch to the foot between the center line of the road and side ditch, $\frac{3}{4}$ inch will be sufficient on a 5 per cent. and $\frac{1}{2}$ inch or even less on a steeper grade. Common sense must govern under each particular condition. The principle involved is to get the storm water into the side ditches with the least current and the least damage to the road. Under no circumstances, however, should the road be flat enough, nor should ruts be permitted deep enough, so as to allow the water to run down the center of the road. A road can be quickly destroyed or badly damaged by the rainfall from a single heavy rain storm running down over the surface, instead of settling away into the side ditches. Water running down a road gathers force as it goes, and gains volume the further it flows, until it becomes destructive. Diverted into ditches prepared for it, the force of the flow is kept off the surface of the road.

In some kinds of earth the bottoms of the side ditches will not cut out much even under a heavy flow of water. Where the ditch reaches to clay, or hard-pan, or hard gravel, or any one of several kinds of hard earth, there is little danger; but if the earth be of a nature which will wash easily, some solid form of bottom must be provided for the ditch on heavy grades. Cobblestones or flat stones, laid carefully in the ditch bottom, curving up at the edges; or field stones set on edge and the spaces filled with smaller stone or gravel, may be used, according to whatever material is most convenient. In some cases it is desirable to put in concrete ditches if the grade is steep enough so that the water will get under those made of stone, and the earth has those qualities which will make it wash out quickly.

An important factor is to get the water out of the ditch before so great a volume and current have accumulated as to cause damage. On most heavy grades an outlet can be found on one side of the road or the other. In such cases the water from the inside ditch should be carried to the outside through culverts at distances ranging from 200 to 400 feet, according to the grade, and discharged from the right of way, together with the water from the outside ditch, into some natural channel of drainage. The culverts should be large enough to carry all possible accumulation of water, and should have pitch enough so that they will keep themselves washed clean of sediment, and carry off the water readily, without backing it up at the opening.

Sometimes, however, it becomes necessary to carry the water down a long hill with no chance to get it off the right of way until the bottom of the hill is reached. In such case the most satisfactory method seems to be to lay sewer pipe, or some other large sized pipe under one of the side ditches, with openings every 200 or 300 feet, with cross culverts from the other side ditch, so that the constantly increasing volume of water may be carried off without damage to either the side ditches or to the road.

Under no circumstances should water-breaks be built across a road. The water should always be carried underneath the road in culverts. These culverts may be made of various forms of galvanized or cast iron, of concrete, or of stone, according to conditions. But the wooden culvert should not be used except under unusual circumstances, as the cost of keeping it up will soon pay for a permanent one.

Summarized, the earth road may be considered as involving:

(1) Width of road and of right of way.

(2) The crown: its height and position on a straight road, on curves, and on side hill roads.

(3) Alignment, location.

(4) Preparing the road under different conditions. Removal of sod, equipment, and methods. Harrowing and rolling.

(5) Exception in certain states where frost is not a factor.

(6) Preparing the side slopes.

(7) Time of year for improving earth roads.

(8) Connection of grade and drainage.

A careful consideration of these factors will make possible the construction of earth roads which, with proper care, should last for years under any ordinary traffic such as earth roads should be expected to stand. When the travel becomes so heavy that a good earth road goes to pieces under it, ordinary economy demands a more substantial surfacing.

Earth Road Maintenance

The most approved method of maintaining an earth road is by dragging. The dragging should begin as soon after the frost is out of the ground as possible, so that the earth is dry enough so that mud will not clog up under the drag, but will smooth out into a smooth surface. The dragging should begin at the sides of the road, going one way and returning the other, and continue by successive trips until the earth is carried to the center of the road, and the crown maintained in its original form.

The dragging should continue until all ruts and deep wheel tracks are filled in and smoothed over, so that the water will run off into the ditches. After every rain the road should be dragged so as to preserve the shape, take the water out of ruts which may have formed, and even up the surface to a proper shape and crown. In some states the law provides for dragging earth roads whenever necessary. Generally this is done by contract with farmers along the road, each having a certain distance to cover, and being compensated, by the mile, for each dragging which may be necessary during the season.

There is a variety of road drags, and road hones, and road smoothers on the market, some made of wood, some of metal, and others of these materials combined. All of them seem to serve the same purpose, though some of the more elaborate ones have advantages in the way of mechanism and cutting facilities which make them useful in reshaping the road after it has been damaged by unusually heavy traffic. Some of these machines are so constructed as to take the place of the road machine up to a certain point where the road has to be reshaped.

The most common and popular form of drag is what is known as the split-log drag. This drag is easily made by anyone with even an elementary knowledge of tools, and its cost is trifling.

Take a log 8 or 10 inches in diameter, 8 feet long, and split or saw it lengthwise in halves. Set the two halves with their flat sides facing the same way, about 3 feet apart, so placed that one end of each shall extend about $1\frac{1}{2}$ feet beyond the end of the other. Connect the two sections of log with two cross-pieces, about 4 feet apart, set in solidly either with a mortice or at least a 3-inch augur hole. On these cross-pieces fasten a board for the driver to stand on whenever it is desirable to give the drag extra weight.

In attaching the hauling chain to the drag it should be so adjusted that the drag will be hauled at an angle which will throw the earth toward the middle of the road, and so that the rear section of log will be directly behind the front one. It is usual, in building a split-log drag, to spike a strip of old wagon tire or other band or strap iron along the bottom or cutting edge of the section, especially the front one, which encounters most of the obstructions. Similar drags may be made of plank by spiking on extra pieces to prevent splitting under the strain.

Constant attention must be given an earth road if it is to be kept in good condition. The stones must be kept off and the side ditches and culverts kept open. Grass or weeds should not be permitted to grow nor sod to form on the roadway. Holes must be filled promptly with fresh earth. If soft or spongy spots develop, the soft earth should be dug out and fresh earth packed into its place.

When a mud-hole is found at a low place in the road, it is not wise to dump stone into it, then cover the stone with earth. Such a course will leave two mud-holes idstead of one, one at either end of the pile of stone.

First, the mud-hole should be drained and some kind of a permanent drain left to carry off the water in the future. Where stone is available a drain of loose stone is advisable. Then fresh earth of a quality which will pack should be put in and the roadway brought a trifle higher than the regular grade, so that when the earth becomes thoroughly firm the grade will be even with that of the adjacent sections of road.

With careful attention to details, and the exercise of good common sense, an earth road carrying any ordinary travel can be kept in good condition the year round except, possibly, in some soils, the few days during which the frost is leaving the ground, at a very moderate expense. But it must be borne in mind that any neglect of a road which permits it to get in bad condition very materially increases the average cost.

After an earth road has been put in good condition the cost of maintenance is comparatively small. But failure to keep up the maintenance is likely to result in as much damage in a single season as proper care would cost in five years.

Earth roads, like all others, require constant and intelligent care and attention.

CHAPTER XI

GRAVEL ROADS

GRAVEL, in some form, is one of the most widely distributed road materials in the United States. Deposits of gravel are found nearly everywhere, except in the alluvial lands in the valleys of rivers and along low sea and lake coasts, and in the higher mountain regions. The scientific theory is that in one of the subdivisions of the Pliocene Age the northern half of the United States east of the Rockies was covered with a layer of glaciers which came down from the north, carrying gravel and other materials, ploughing the surface of the country into irregular shapes, and in their melting depositing the gravel and clay and stone where the glaciers which held them finally rested. The escape of the waters formed by the melting of the glaciers is supposed to account for the location of the channels of rivers, creeks, and smaller streams as the waters sought sea level.

The scientific theory of the glacial agency in accounting for the presence of gravel is confirmed to a certain extent by the fact that certain gravels in some sections are plainly identified as particles of certain rocks which are found in regions far north. But other gravels, in localities not supposed to have been affected by the glacial movement, may have been the product of a previous "Geologic Age," the status of which is not well defined. Students who devote their lives to the various theories are able, in some instances, to furnish practical information. Their researches and the results are entitled to respectful consideration.

Gravel that is found in pits or beds is generally considered better for road building purposes than that found in the bottoms of creeks or rivers. The reason for this is that most pit gravel has a thin coating of clayey substance covering each particular particle or pebble of which the gravel is composed. Gravel with such a coating constitutes what is known as "cementitious" gravel; that is, gravel which will compact into a firm, solid mass under a road roller or under traffic. Gravel taken from the beds of streams does not have this coating. It has been washed off by the action of water, and the material has become what is known as "washed gravel." If a binder is used in building the road surface the washed gravel is far superior to the "cementitious" gravel. In fact, if a bituminous binder is used in a gravel road it is necessary that the gravel be clean of coating material. If pit gravel be used it must be artificially washed to remove the clay or other coating on the particles.

In building a gravel road the earth should be excavated to a distance of 8 inches below the level of the finished road and firm shoulders left at each side. The subgrade thus formed should have the same curve or crown as the road is expected to have when completed. This subgrade should be rolled with a machine roller, the heavier the better, and if any soft or springy or spongy spots develop under the rolling, the earth should be dug out and solid earth which will pack put in its place. When wet spots are found they should be provided with underdrainage. (See Chapter IV.)

There are two methods of placing the gravel. One is to deposit the gravel to a depth of 10 inches and then rake the small pebbles to the top, allowing the larger ones to be worked to the bottom. Under a roller the 10 inches of loose gravel will compact to about 8 inches in the finished road. It is considered by most practical road builders, however, that it is better to lay the gravel in two courses. The bottom course should be of the larger pebbles, but not larger than $2\frac{1}{2}$ or 3 inches in diameter, and not smaller than about $1\frac{1}{4}$ inches. This course should be laid about 6 inches deep, a little sand from the screen spread over it, and rolled to a thickness of about 5 inches or perhaps a trifle less. On this should be spread 4 inches of properly graded finer gravel ranging in size of particles from sand to 1- or $1\frac{1}{4}$-inch pebbles, and the whole mass rolled to a thickness of 8 inches. In such a road the "pit" or cementitious gravel should always be used, and the rolling should be thorough. The rolling should begin at the sides and work toward the center, making a smooth even finish on the surface and uniform hardness below.

The old-time method of dumping gravel on the roadway and letting the traffic "wear it down" is neither economical nor satisfactory. It is a waste of material, and even if the road be in a locality where gravel is plenty and cheap the hauling costs money; and such a road can never give satisfaction because the larger pebbles will always be working to the surface, and the road will be rough and uneven. The same amount of money, if put into proper construction, will make a much more satisfactory road.

Probably the gravel road has reached its highest modern point of excellence in the states of Vermont and New Hampshire. These states, with limited means at their disposal, have gone forward year by year, putting in stretches of gravel and other roads where they were the most needed, until both states have long stretches of excellent roads—mostly gravel.

The maintenance of these gravel roads is simple. It consists in spreading a thin layer of fine gravel over the surface and rolling it, if a roller be available. If not, the traffic will compact it by reason of the fineness of the material and the thinness of the layer. In some

Fig. 19.—A gravel road in New York State.

sections of the country gravel roads require dragging at frequent intervals. This may be due in some cases to the nature of the coating or cementing factor in the gravel itself; or to unusually heavy traffic; or to a travel that "tracks" until ruts are produced; or to extremes of wet or dry weather. When the winter weather is so severe and the drainage so insufficient that the subgrade becomes honeycombed, traffic should be kept off the road during the thawing season to prevent its destruction, and the road should be rolled with a heavy machine roller as soon as the frost is out of the ground. Where uneven places appear more gravel should be added to fill depressions until the road surface is smooth and even. These conditions are likely to develop in but few sections of the country, and only in occasional seasons of especially low temperature, or where proper attention to drainage has not been given.

In earlier days Ohio and Indiana built many miles of gravel roads as "feeders," or connections, of the old National Road. What these states did systematically many other states did sectionally. These states each built several thousand miles of gravel roads, mostly without careful attention to grade or to the other features which modern roads require. These roads, under the older conditions of travel and traffic, were an important factor in the development of the agricultural and other interests of those sections at the time they were built. Proper maintenance, however, was in most instances neglected, and modern traffic finds a large percentage of those old gravel roads in a practically ruined condition.

In many instances, however, these old gravel roads furnish an excellent foundation for a new road, and much money is saved by making use of them. It is often found that the dumping of fresh gravel on the road from year to year to be packed down by the traffic has resulted in a foundation solid enough and deep enough to carry any kind of a road surface. In such cases the foundation should not be disturbed. The surface should be cut down with a scarifier until the road has the proper or desired shape, defective places filled in and rolled or tamped, and a surface put on of the gravel, as described previously. Of course, it may be necessary at times, in order to reduce grades, widen curves, or for other local reasons, to dig up the old foundations; but the necessity should be carefully studied before such a course is decided upon.

There is a great variation in the quality of gravels. In some the pebbles are round or nearly so. In others they are of irregular, angular shapes. The angular shape makes a better road, as it compacts better. The pebbles of which some gravels are composed are of trap rock, others are of different colors of granite, others of quartz, and still others of softer stones. What is known as "blue" gravel is usually of trap rock and is held by most road builders to be the highest grade of gravel for road purposes; but those of granite and quartz, if hard, make good road surfaces. Almost any gravel which will pack well makes a good lower, or foundation, course.

In several of the Southern States a very soft gravel is found in liberal quantities. Owing to its softness it does not make a durable or satisfactory road surface, grinding into dust during dry weather. But it packs excellently and makes a first-class foundation for whatever surface the traffic may require.

In building a gravel road special attention should be given to grading the gravel, especially in the surface course. The pebbles should be in the proper proportion of the various sizes so that the smaller ones will fit into the spaces between the larger, and sand enough should be added to fill the spaces between the smaller stones. When gravel is screened to get the proper proportion of sizes the tailings from the screen may be used instead of sand, as they usually consist of the clayey or other cementing substance forming the coating on the pebbles of which the gravel consists.

Gravel should not be dumped on the roadway before spreading. When spreading wagons are available their use is an advantage. When they are not available the gravel should be spread with shovels from wagons, or dumped on boards and spread with shovels, so as to make an even surface. Otherwise the road will wear unevenly.

When the traffic on a gravel road becomes so heavy that the dust becomes a nuisance, it requires a bituminous treatment on or in the surface course. This takes it into the class of Road Surfaces, and is described in Chapter VI.

In some sections of the country large numbers of boulders are found of the same quality of stone as the gravel in those localities, and sometimes imbedded in the same pit. These boulders are usually very hard. They range in size from the gravel up to several feet in diameter. Some roads have been built by using such of these boulders as range from 3 to about 12 inches in diameter as foundations. This is not advised, as the irregular character of the foundation will naturally show in equally irregular wear on the surface. It is better to break up these boulders, crush them in a machine crusher, screen the stones into regular sizes, and use them as broken stone. Their quality is such that they usually make most excellent roads. This method has been pursued quite extensively and very successfully in parts of Wisconsin, and to a lesser extent in many other localities.

CHAPTER XII

SAND-CLAY ROADS

No road offers a greater opportunity for the exercise of common sense or "horse sense" in its construction than the sand-clay or clay-sand road. This is due to the fact that the excellence of the road depends on the proper mixing of the clay and sand, and there are so many varieties of clay and so much difference in sand, even in the same vicinity, that the most careful attention is at all times necessary.

The theory of the sand-clay road is that the sand furnishes the body or carrying power of the road, while the clay supplies the "sticking" power which cements the particles of sand together, and holds them in a solid mass; consequently, the coarser the sand, the more clay required to fill the spaces between the particles of sand. With very fine sand the amount of clay required is very much less. Some road builders suggest the following plan for determining the amount or proportion of clay necessary in the mixture:

"Take a measured quantity of the sand to be used—say 1 cubic foot or ½ cubic foot—have it thoroughly dry, and place it in a small tub, or water-tight box, or large bucket. Then put in water slowly, so it will settle all through the sand, and come just to the top of the sand. Then drain off the water and carefully measure its bulk. The amount of water, in bulk, is exactly the amount of clay which should be mixed with the sand to fill the spaces and stick the particles of sand together."

While this method will establish the proportions of sand and clay to be used, there is so much difference in clays that a careful study must be made of them to determine their fitness. Not all clays make satisfactory sand-clay roads.

Some clays are very sticky and have excellent binding power, holding the grains of sand firmly together. Others have little or no binding power, due generally to particles of mica in the clay, and are almost useless for road purposes. Some clays will absorb but little water; others are almost like sponge. Some clays will keep their shape and sticky qualities for a long time in water; others will melt down and dissolve almost immediately when submerged in water.

The clays which are the most sticky and which resist the dissolving action of water the most make the best roads. The best way to determine whether a clay is suitable for road purposes is by actual test. If clay sticks to the hands and fingers when they are wet, it is likely to be sticky enough for road purposes. If a ball of clay holds its shape for a number of hours when immersed in a basin of water, it is likely to hold its position when mixed with sand in a road. If it dissolves easily it should not be used in road work, as the first rain of any considerable duration will dissolve it and leave only a sand road.

Clay from each particular bed or deposit should be tested before using, as frequently different qualities of clay are found in close proximity to each other. For the benefit of those who have the facilities and desire to make chemical tests the following technical definition of clay is given:

"Clay consists of hydrated silicate of aluminum, with small proportions of the silicates of iron, calcium, magnesium, potassium, and sodium. Its tenacity and ductility when moist and its hardness when dry varies in proportion to the variance in the quantities of the materials of which it is composed."

If clay stands up well in the banks of running streams, and does not wash out readily in stream bottoms, it should be a good road clay. When it cuts badly it should be used with great caution, if used at all, in road building.

In building a sand-clay road on a sand subsoil the sand road should first be graded and smoothed, and provided with side ditches. Then clay should be applied to the top to a thickness of 6 inches and ploughed in. At first the plough should not turn up more than 3 or 4 inches of the sand. Deeper ploughing can follow

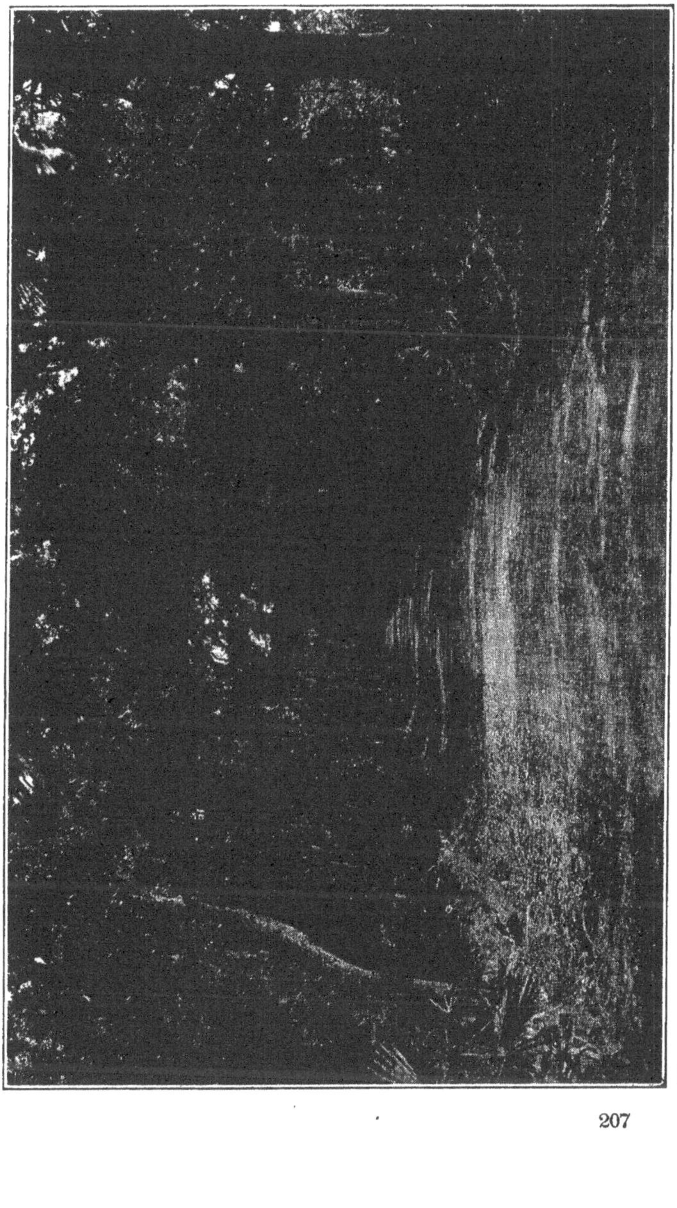

(Photo by Chapin, Jacksonville, Fla.)

if more sand is necessary in the mixture. Then disk harrows should be used and water applied so as to first completely pulverize and then puddle the mass, after which it should be left to dry out, with the traffic kept off. This is often a difficult matter, as in certain conditions of weather it may take some weeks for the sticky mass to dry. When it is dry enough it is well to use a road roller on it if one be available; in any case, it should be trimmed down with a road machine, and travel let on by degrees, light vehicles at first, so that the surface may be packed gradually.

When the subsoil is of clay the surface should be ploughed to a depth of about 4 inches, thoroughly pulverized, dry, with a disk harrow or with whatever utensils will do the work. The clay, in this pulverized form, should be left to dry out thoroughly. Then about 8 inches of clean sand should be placed on top, and mixing proceed with the disk harrows, or by hand mixing with shovels, or by any other means which will secure a thorough mixing of the materials. Then water should be applied in considerable quantity and the mixing continued until the mixed mass is saturated. It should then be left to dry out, as in the case where the clay was placed on the sand, and the subsequent treatment should be the same.

Maintenance of sand-clay roads must begin as soon as the road is opened to travel. The utmost care in mixing the materials will not always insure a uniform surface, and sand or clay must be applied as needed until the road wears to a uniform surface. After each rain or spell of heavy weather a sand-clay road should be gone over and defective places repaired. Often dragging with a split-log or other drag is of benefit, with the addition of clay or sand as required.

Practice varies among the advocates and builders of sand-clay roads as to the degree of curve or crown which such roads should have. Some builders insist that the roads, especially when on a sand subsoil, should be flat, so that all the moisture except excessive rainfall will settle into the sand and add to its firmness. Others contend for a high crown, so that the water will run off quickly.

It is probable that the shape, to be of the most value to the traffic, should neither be too rounded nor too flat; but must depend much on the particular quality of the clay in the road.

North Carolina, South Carolina, and Georgia have built sand-clay roads extensively, and in most cases they have given excellent satisfaction. There has been a wide difference in the cost of these roads, ranging from $250 to $2000 a mile, according to the proximity and cost of materials, the cost of labor, and other factors.

Heavy loads on steel tires are reported to be exceedingly damaging to sand-clay roads. Light buggy travel to almost any extent does not seem to injure them, while automobile travel, especially on pneumatic tires, is stated to be a positive advantage, as the tires keep the surface of the road smooth and well and evenly packed.

In some sections of the country sand and clay are found in a natural mixture. Where this appears in the roadway all that is necessay is to remove the soil covering, and properly grade and drain the road. In other cases the soil should be removed to a depth of 10 or 12 inches and the natural mixture dumped into its place and properly shaped, preferably with a road machine, and, if possible, rolled with a power roller. This natural sand clay should be thoroughly packed without the addition of much water, though a little may be necessary.

CHAPTER XIII

TOP-SOIL ROADS

As the name indicates, "top-soil" roads are built mainly of the surface soil from the fields adjacent to the roadway. This soil is placed on the subgrade after the road has been properly graded, shaped and drained, and prepared for the reception of this material.

The top-soil roads range in character all the way from a sand-clay road to a road of cementitious gravel. The top soil which is used for road purposes is composed principally of sand, gravel, clay, and a small proportion of silt, with some decayed vegetable matter included. In the sections where this top soil is found in a proper mixture for road purposes the subgrade is usually of sand or gravel; or, if there are spots where the sand or gravel does not appear, there are gravel pits near enough to make a sand-gravel subgrade by a short haul of the material. It is held by those who have given the top-soil road the most careful study that under favorable conditions it is the cheapest and yet the most durable and satisfactory road that can be built.

The top-soil road has reached its highest development, apparently, in the state of Georgia, though many miles have been built elsewhere. Reports from the work in that state are fragmentary by reason of the fact that Georgia has no state highway organization, the only practical information obtainable being through personal investigation by the author, and occasional scientific addresses or reports on the subject by members of the faculty of the University of Georgia.

Practical experience has shown that while the sand-clay road, described in Chapter XII, requires a careful selection of sand and clay for its success, the top-soil road may be successful with a wider range of materials, amounting at times almost to a reversal of the principles involved. For instance, a clay which would not be considered in building a sand-clay road, may produce the most satisfactory results when found in a natural combination in a surface soil. And silt and decayed vegetable matter, which would not be admitted in a sand-clay road, become factors of excellence in their natural combination in the top soil. These facts show that nature can prepare a material which cannot be duplicated successfully by man.

A scientific discussion of this subject is not within the province of this book. The fact that the materials are available in many sections of the country justifies a reference to the Engineering College of the University of Georgia for technical information by those who may desire it.

Practically, the top-soil road is a combination of

earth, gravel, and sand-clay formed into one, with nature performing most or all of the mixing. Where the materials are found in the proper proportions the

Fig. 21.—Top-soil road in Alabama.

results are very satisfactory; where there is too much or too little of any one of the materials, as of clay, or sand, or gravel, they may be added, though care must be taken not to add too much.

The result, when the proper combination of materials is found or prepared, is a good hard road, which produces little dust, does not rut under ordinary heavy loads, and wears for several seasons with but little maintenance charge. In some cases the road becomes almost as hard as concrete, though not so rigid. At times its hardness will resist the cutter of a road machine, especially in dry weather.

In building a top-soil road the top-soil material is deposited to a depth of 10 or 12 inches. If the road be of a given width the subgrade should be formed by excavating, and shoulders made. Either the work should be done in wet weather or a reasonable amount of water should be applied, so that the material may be compacted thoroughly with a roller or with traffic, and so distributed that the larger gravel pebbles shall be at the bottom. No pebbles larger than 3 inches in diameter should be allowed in the road.

So far as is known to the author there are no examples of top-soil roads in the Northern or Central States. There seems no reason, however, why such roads should not be built in any locality where the soil and subgrade offer the proper conditions. The questions of drainage and protection against frost action are practically the same as with earth roads.

The cost of top-soil roads is given as approximating an average of $1200 per mile.

CHAPTER XIV

MACADAM ROADS

"MACADAM" is the name applied to almost every variety of road the surface of which is made of broken stone. How the name of John Louden McAdam, who built broken stone roads in England, 1815 to 1824, became attached to this type of road in the United States it is difficult to determine. France had been building the same type of roads, instituted by the Engineer Tresaguet, for more than sixty years before Mr. McAdam became known as a road builder, and had constructed more than 15,000 miles of national roads under the system. In the United States several hundred miles of the Old National Road, in some localities called the Old Cumberland Road, had been built of broken stone in Maryland, Pennsylvania, and Virginia before McAdam's construction in the north of England became noted.

Notwithstanding these facts of priority, the name, corrupted to "macadam," has been used generally in the United States in its application to roads of this character. Efforts of scientific organizations to use another name have failed, so far as the general public is concerned.

Until the automobile became the dominant factor in road travel macadam roads were considered the best type of roads not only for the country, but also for villages, and even for the residence streets in some large cities. At the present time macadam without some special surfacing is hardly considered for main roads in thickly settled communities, and especially where the motor traffic is heavy.

But there are many thousands of miles of road in the country where broken stone roads can be economically built and maintained. Where the average travel does not exceed 250 or 300 vehicles per day, about half automobiles and one-quarter heavily loaded wagons on steel tires, the macadam road will answer every purpose. When the traffic gets heavier than that, special surfacing should be provided for.

The width of a macadam road is a matter for local judgment in connection with the traffic to be carried and the amount of money per mile available. In some cases roads have been built only 8 or 9 feet wide. These were in sections where the heavy loads are mostly hauled in one direction, and a good earth road is usually built alongside to accommodate the light travel going the other way. Under such circumstances these narrow roads have often served a useful purpose.

For general team traffic a road should not be less than 12 feet wide, and where there is automobile travel the width should be at least 15 feet; 16 feet is still better, as the safety of the vehicles is better provided for and the edges of the stone portion of the road are not so likely to be damaged by wheels running off and on, and grinding along the edge.

After the proper grade and drainage have been provided, as stated in the chapters on those subjects, the earth should be excavated to a depth of 8 inches, leaving a firm shoulder on each side. The subgrade thus formed should have the same shape and crown as the surface of the completed road, and should be rolled with a heavy machine roller of at least 10 tons' weight. It is necessary that the subgrade be packed hard, both for the purpose of giving an even support to the road, and to prevent the stones from settling or being driven into the subgrade. The stone is the most expensive part of the road and that which would get into the subgrade would be wasted.

In former years it was the custom to deposit the stone in one course. Under modern conditions of travel this method has been found unwise and has been generally abandoned. Practically all macadam roads in recent years have been built in two courses, the bottom course about 5 inches thick when rolled, and the top course about 3 inches, making 8 inches

Fig. 22.—Completing rolling of subgrade for macadam road.

Fig. 23.—Rolling the foundation course of stone for a macadam road.

in all. A road built in this manner wears more evenly.

Practically all stone crushers have revolving screens which separate the various sizes of stones and deposit

Fig. 24.—Completed macadam road (New York construction).

each size in a separate bin. The stones in the lower course of a macadam road should range from $1\frac{1}{4}$ to $2\frac{1}{2}$ inches in size and spread 6 inches deep on the subgrade. This will compact to about 5 inches under the roller.

The stone should not be dumped on the subgrade.

Fig. 25.—Macadam road with asphaltic binder (Ohio construction).

It should be spread with an automatic spreading wagon or spread from the wagons with shovels. If it is necessary to dump the stone, it should be on dumping boards and then spread with shovels. No earth or loam or sand should be placed on the lower course, nor is it necessary to apply any water. The theory of the broken stone road is that the stones are so packed together as to make a solid structure which will carry any reasonable weight which may be placed upon it.

On this lower course is spread 4 inches of smaller stones ranging from $\frac{1}{2}$ to $1\frac{1}{4}$ inches in size. This is leveled to the proper shape with rakes, so that the smaller stones will be at the top. Then the rolling should begin.

Rolling the top course should begin at the sides, one side immediately following the other, with the roller extending over the earth shoulder, so as to compress the earth and stone together. By doing this the sides of the road will be made firm. Then the rolling should continue from the sides toward the center.

As the rolling progresses water should be applied and stone chips or screenings spread on the surface. Small fragments of stone and stone dust are formed under the action of the roller and by the crushing of the edges of the stones as they are forced into permanent places. These fragments and dust must be washed down into the mass by the water and form the cementing binder which holds the stones in place. Sand should never be used under any circumstances. The water may be applied by a hose with a spray nozzle or by a sprinkling wagon which will deliver the water freely. The watering and rolling should go on at the same time and continue until the water will no longer settle into the road, but will run off at the sides; then the rolling should be kept up until the surface is hard, so that the roller will not make an impression on it. Then the road is ready for traffic.

While the foregoing description covers the construction of a regular two-course macadam road, there are several modifications which may be made, especially as regards the bottom course, to save expense.

Naturally, as the wear comes on the top course, the stone for this course must be of the best quality obtainable at anything like a reasonable cost. Almost any kind of local stone will answer for a lower or foundation course. Even coarse gravel is sometimes used, with the top course of broken stone; and many of these roads have given excellent satisfaction.

Roads have been built in some localities where traffic is light by using 8 or 10 inches of ordinary field stone, without crushing, for the lower course. These stones were thrown loose into the subgrade, and were broken up by men with sledges, so that no stone should be more than 8 inches in its largest dimension. Then the mass was rolled until every piece of stone found a firm resting place, after which the regular macadam top course was put on, as previously described. Some of these roads have worn for several years with good results. They cost much less than a regular macadam road.

Telford is rather the name of a foundation or lower course than of a road. The name is that of an engineer who built roads in the south of England and in Wales at about the same period that McAdam was building them in the northern section. In all modern practice where Telford's method is used it is as a foundation, with a macadam top course.

In preparing the subgrade for a Telford foundation the same method should be followed as for a macadam road, except that the excavation should be deeper and the subgrade need not be so thoroughly rolled. The same care should be given to grade and drainage and to having good, substantial shoulders.

Stones ranging from 6 to 12 inches in length and 2 to 4 inches in thickness are set up edgewise in courses, crosswide of the road. The largest and flattest edge is set downward on the subgrade, and the stones placed as close together as possible, so that the flat edges cover the subgrade as nearly as possible, and so that the stones in place may resemble a well-built wall, with the face downward, and the irregular points of the stones sticking up. Then, with a sledge-hammer, the points are broken off, and the pieces driven down between the stones to wedge them tight into their places. The breaking and wedging of the upstanding points leaves all the stones at about the same height, and the wedging makes the structure of the foundation strong enough to distribute over a considerable space any weight which may come on the road. For this reason the Telford foundation is often used in soft and spongy and springy soils. In some localities where stone is plenty it has been found economical to use it regularly. While more expensive than macadam foundations, some veteran road builders hold that it will carry a heavy traffic as well as a still more expensive one, such as concrete.

On the Telford foundation is placed the macadam top course, as previously described. It is a common custom with some road officials to use a regular macadam foundation where the subsoil is hard and solid, and to put in Telford wherever soft places appear, even a rod or so in length.

The quality of stone in the surface, or top course, is of the greatest importance. In many localities stone suitable for the bottom or foundation course is found in ample quantities, but the surfacing stone must be purchased and shipped in. Generally the estimates for the stone on the road are made by the cubic yard. For instance, 1 cubic yard of stone would cover $1\frac{4}{5}$ linear yards (a little more than $5\frac{1}{3}$ feet) of 15-foot road 4 inches thick. But such stone, whether or not purchased by the cubic yard, always has the freight charges figured by weight. It is also to be noted that crushed stone, shipped in freight cars, will always settle down in transit, so that if purchased by the cubic yard it is always better to have an understanding as to whether the measurement is to be taken at the place of shipment, when loaded on the cars, or at the point of delivery, before the cars are unloaded.

The weight of stone varies so greatly that there is no given rule that can be applied. The only safe way is to weigh a measured cubic yard of the particular stone. Some crushed stone weighs but little over a ton per cubic yard, while other kinds of stone weigh nearly 1½ tons. And the weight of the stone is not to be depended on as a guide to its quality for road surfaces. For instance, some crushed granites weigh as low as 2200 pounds per cubic yard; but granite is so variable in quality that it may make an excellent and durable road surface, or it may wear out in one or two seasons.

On the wearing quality of the surface stone the long or short life of the road depends. In some sections of the country trap-rock or other high-grade stone is available. In other sections, where the stone has to be shipped some distance, the cost of freight and the cost of the stone from various points must be taken into consideration, together with tests of the stone for road surface purposes.

The United States Office of Public Roads and Rural Engineering, Department of Agriculture, Washington, D. C., has a large and well-equipped laboratory for testing road materials. The tests are made free of charge at the request of any road official. The official asking for the test, however, must pay the express charges on the samples sent for that purpose. The report gives the hardness, the toughness, the crushing resistance, the binding qualities, and other properties of the stone which make for or against its value as a road surfacing material. Some state highway departments and some state universities have similar facilities. The use of these facilities to the fullest extent is recommended in the selection of stone for the top course of macadam roads. On this wearing surface depends largely the economy or the waste of money involved in building the road.

Many quarries and stone companies which make a business of producing stone for road purposes equip their salesmen and representatives with copies of the government or state official tests of their stone. With entirely reliable concerns these reports may be depended on. But cases are not unknown where inferior grades of stone have been supplied, where the best was expected. In these cases it is charitable to suppose that the stone manufacturer inadvertently permitted the stone to be shipped without proper examination or supervision. It is usually safer to guard against mistakes of this kind by having tests made when the stone arrives, or if previous tests have been made, comparing the stone very carefully with the sample tested.

The following figures are rearranged for practical application from an elaborate set of tables worked out by Mr. R. A. Meeker, State Highway Engineer of New Jersey. In that state contracts for macadam roads were let on a basis of weight of stone instead of per cubic yard, as is the custom in most states. The crushed stone used on New Jersey roads weighs from 2350 to 2500 pounds to the cubic yard.

SQUARE YARDS IN 1 MILE OF ROAD

8 feet wide	4,693⅓	square yards
12 " "	7,040	" "
14 " "	8,213⅓	" "
16 " "	9,386⅔	" "
18 " "	10,560	" "

Any other width can easily be computed from these figures.

The following table contemplates 6 inches of foundation and 4 inches of surface of loose stone, which will compact to 8 inches under proper rolling; and stones weighing about 2400 pounds per cubic yard:

STONE REQUIRED PER MILE

Width of road.	Depth of stone.	Amount per mile.
8 feet	4 inches	875 tons
8 "	6 "	$1312\frac{1}{2}$ "
10 "	4 "	$1093\frac{3}{4}$ "
10 "	6	$1640\frac{5}{8}$ "
12 "	4 "	$1312\frac{1}{2}$
12 "	6 "	$1968\frac{3}{4}$
14 "	4 "	$1531\frac{1}{4}$
14 "	6 "	$2296\frac{7}{8}$ "
15 "	4	$1640\frac{5}{8}$ "
15 "	6 "	$2460\frac{15}{16}$ "
16 "	4 "	1750 "
16 "	6 "	2625 "
18 "	4 "	$1968\frac{3}{4}$ "
18 "	6 "	$2953\frac{1}{8}$ "

With these figures, when the cost of the stone per ton, or its cost and weight per cubic yard, is ascertained, the cost of the stone per mile for the two courses can be easily figured. For any width of road not named, or any depth of stone other than those given, the computation is one of simple arithmetic.

In figuring stone by the cubic yard, 1 cubic yard will cover 9 square yards of road to a depth of 4 inches, and 6 square yards to a depth of 6 inches.

Repairing Macadam Roads

When a macadam road ravels or when holes develop in the surface, the bad place should be cleared of loose material and the dust carefully swept out. It is usually well to dig a trifle into the edges and bottom of the defective place to be sure of getting to firm material. Then fresh stone should be put in, thoroughly watered, and tamped or rolled into place. It is not wise to let a bad spot develop until it is large enough to call for the use of a roller. A tamper, such as is used by pavers in cities, should be kept ready for use in repairing the smallest defect in the road, and piles of crushed stone should be kept within wheelbarrow distance at all times. Holes and ravel spots develop very rapidly when once the surface is broken and should be repaired at the earliest possible moment.

Maintenance, as distinguished from repair, consists of keeping the road surface so protected that the stone structure of the road will not wear out or otherwise become permanently damaged. This maintenance consists in sweeping the road surface carefully so that it is free from dust and other matter, and spreading about $\frac{1}{2}$ or $\frac{3}{4}$ inch of fine stone ranging from screenings to $\frac{3}{8}$ inch, or possibly $\frac{1}{2}$ inch in size. This may be spread over the entire surface or over the worn places, as the conditions may suggest, and the work should be done at the beginning of the wet season. The traffic will then pack the stone into place. Roads have been maintained in this manner for many years.

It must always be borne in mind that the surface of a macadam road is to be kept waterproof. Any wear or damage, or opening of any kind that will let the water into or through it is likely to cause extensive and expensive breaks.

Bituminous Macadam Roads

About 1906 and 1907 the growth of the automobile traffic on the main roads caused so much wear that strenuous efforts were put forth to find a remedy and a prevention. Investigation of the subject showed that the theory of the macadam or broken stone road would not hold under the new conditions.

The theory of the macadam road, as previously stated, is that the particles of stone and dust broken from the stones by the steel shoes of horses and the steel tires of vehicles are driven into the crevices between the stones and, aided by water, form a binder which makes the road waterproof and holds the stones in place. When the automobile came this was changed. The low-bodied, swift-moving vehicle with pneumatic tires sucked the binder out from between the stones and threw it in the air. Then more steel-shod hoofs and steel tires would break down more stone, to be thrown off by following automobiles, and the road was soon destroyed.

Experiments by nations and states show that a proper bituminous material mixed with the top course of stone prevents this extraordinary wear. The situation was so grave that International Road Congresses were held in Paris in 1908, in Brussels in 1910, and in London in 1913 to discuss road subjects; primarily the protection of the broken stone roads.

In the use of bituminous materials such as asphalts and tars the changes of practice have been numerous and rapid. Two methods of applying the bituminous material were developed almost from the beginning. One was the "penetration" method, and the other the "mixing" method. Both have so far been improved by modern road builders that only the roads built by the penetration method may now be classed as "bituminous macadam." Roads built by the improvement of what was called the "mixing" method are more properly classed as "bituminous concrete" and are treated in Chapter XVII, under "Bituminous Roads."

In the use of a bituminous material for binder in a macadam road the practice and results have been variable. In England the most satisfactory results have been obtained by the use of tar. French and Spanish road builders have been unable to obtain the same degree of success with tars of any kind, but have found that asphaltic preparations give the best results. In the United States both tar and asphalt have been used extensively, both in experiments and in practical construction.

The experience of the last few years seems to show that in most sections of this country asphalt has the greater and longer-lived holding power. So prominently has this fact been demonstrated that in modern specifications "bituminous" macadam is rarely named; the usual term being "asphaltic" macadam.

The foundation for an asphaltic macadam road is the same as for a macadam road, except that it may be a little thicker, to make up for a thinner surface course. On the foundation course, before its final rolling, should be spread a thin layer of loam or other available earth. This is to prevent the asphalt, when poured into the surface course, from being wasted by dripping down among the foundation stones where it would do no good.

The surfacing stone, ranging from ½ inch to 1¼ inches in size, is spread on the foundation course and carefully raked over to a depth of 3 inches of loose stone. Sometimes this is gone over once with a roller to "key" the stones in place; that is, to get each particular stone set lightly into the place where subsequent rolling will fix it firmly.

On and into this stone is poured 1½ gallons of hot asphalt. The asphalt is usually heated in wheeled kettles along the roadside, and the pouring done by hand from vessels specially designed for that purpose.

When the asphalt is applied the stone should be thoroughly dry. Closely following the pouring of the asphalt, shovelers will spread a thin coat of screenings, and a heavy roller—at least 10 tons—should be close behind. The roller should never be more than 15 to 30 feet behind the pourers, with the men spreading screenings between.

After rolling, another application of hot asphalt is made of ½ gallon per square yard, and covered with ⅜ to ½ inch of screenings. Then the rolling continues until the surface is hard and smooth and until the heavy roller will make no impression on it. Then the road may be opened to traffic.

Experience in many sections has shown that this last ½ gallon of asphalt and the last coat of screenings form a carpet or wearing surface that perfectly protects the structure of the road; and that a similar application, made once in two or three years, continues the life and excellence of the road indefinitely.

Asphalts for this purpose are of different grades and prices. As a general rule the best and most durable material costs the most at the point of shipment. (See Chapter XVII on Bituminous Roads.)

CHAPTER XV

BRICK ROADS

THE general use of brick as a surfacing material for country roads is usually confined to those main roads where the travel is exceptionally heavy, or where the cost, by reason of the nearness of the brick manufacturing plants, or the absence of other local road materials, places it in competition with other types of roads. Brick is not a "cheap" road-surfacing material in any sense of the term. It is neither low in first cost, nor is it subject, if properly constructed, to heavy charges for upkeep nor to early destruction. In other words, brick roads, when properly built of good brick, while expensive to begin with, last a long time, cost little to maintain, and give general satisfaction. In all cases, however, much care must be exercised in the selection of the brick, the preparation of the foundation, and the other details of construction.

Brick roads and pavements were used in Holland several hundred years ago, but they do not seem to have been in use to any considerable extent elsewhere in Europe. A number of isolated cases of brick pavements in this country at a comparatively early date may probably be traced to early Dutch colonists. It is only within the last thirty or forty years that brick has become a standard paving material in cities, and less than half that time since is has been employed to any considerable extent on country roads.

Paving brick varies greatly in color, ranging through several shades of buff, brown, and red, and some are found of a salmon color. The color, however, has little or nothing to do with the quality of the brick or with its suitability for paving purposes. The bricks are made from shales or from fire clays, one or the other of which is found in many sections of the country. Shales are chemically of the same general composition as some kinds of clay; but they have become hardened and settled into thin layers something like slate, and are found in large deposits. They are made up mostly of silica, alumina, oxide or carbonate of iron, with traces of lime, magnesia, and other ingredients. Various fire clays and some potters' clays of coarse quality possess practically the same ingredients; but surface clays are not often found which will make good paving brick.

Paving brick must be hard, tough, and dense. These qualities are produced by having the proper clay, the proper mixture, and the proper burning. A lack in any one of these three factors may produce a brick which has some, but not all, of the three necessary qualifications.

Tests are applied to paving bricks in three ways: What is known as the "rattler test," to determine the hardness; the "absorption test," to determine the density; and the "cross-breaking test," to determine its toughness.

The rattler test consists of putting 10 bricks, which have been carefully weighed, into a revolving iron barrel specially prepared, and which has been adopted by practically all scientific bodies, by many states, and by the United States Office of Public Roads and Rural Engineering as the "standard rattler." With the 10 bricks are placed a number of round steel shot $3\frac{3}{4}$ inches in diameter, weighing about $7\frac{1}{2}$ pounds each, and some others of one-half that diameter, weighing a trifle less than a pound. The barrel is then revolved at the rate of about thirty revolutions per minute, one revolution in two seconds, until it has had 1800 revolutions. Then the bricks are carefully weighed, and the loss in weight noted. In some bricks of high quality the loss in weight is as low as 16 per cent.; sometimes with a brick which looks to be good the loss runs as high as 40 per cent. The American Society for Municipal Improvements requires that for heavy traffic bricks shall not lose more than 22 per cent. in weight in the rattler test; for medium traffic, 26 per cent.; and for light traffic, 28 per cent.

Samples for testing should be taken from each carload of brick, and should include the softest, the medium, and the hardest burned. Generally, if the tests show that there are but few soft bricks, the contractor is permitted to cull them when unloading from the cars; if the proportion of soft bricks is large, the whole carload should be rejected.

The absorption test consists of placing a carefully weighed brick in water, letting it soak for forty-eight hours, and then weighing it again. This will establish the density by figuring the amount of water it has absorbed. In sections of the country where freezing need not be considered, the amount of water absorbed makes little difference. In cold climates, where the road surface becomes frozen hard every winter, the less water the brick will hold, the longer will be the life of the road.

The cross-breaking test consists in setting a brick up edgewise, the same as it is laid in a pavement, on supports 6 inches apart, and applying the breaking load in the center. Many road builders omit this test if the rattler test is satisfactory.

Bricks for road purposes are usually $3\frac{1}{4}$ inches wide, $8\frac{1}{2}$ inches long, and 4 inches high. That is, a brick $3\frac{1}{4} \times 8\frac{1}{2} \times 4$ inches is set on edge crosswise of the road, except at road crossings, where they are set diagonally

(Photo by Chapin, Jacksonville, Fla.)

Fig. 26.—Brick section of Orange Park road in Duval County, Florida. Bricks are laid on sand foundation.

or cornerwise, so as to present a crosswise front at each road approaching the crossing.

The brick is burned very hard, until it becomes vitrified; that is, given a very hard and usually a glazed surface. This burning usually requires seven to ten days' time and 1200 to 2000 degrees of heat to secure the proper hardness and the right degree of vitrification. Then a period of about as much more time is necessary to permit the brick to "anneal" and cool off gradually without cracking or other damage.

Two kinds of road brick are in general use: the "wire-cut lug" brick and the "repressed" brick. The wire-cut lug bricks are formed by their being cut off from the prepared bar of stiff clay by wires running in grooves, so that lugs are made on the sides which hold the bricks at a proper distance apart to receive the filler when they are placed in the road. Repressed brick are cut off the bar of clay by an automatic cutter, and each brick placed in a die where the corners are rounded and the brick is compressed a trifle in size. In the repressing some manufacturers leave the name or factory mark raised so as to prevent the bricks from lying too close together in the roadway, thus serving a purpose similar to that of the lugs.

The general feeling among a large number of road builders who have been personally interviewed is that the wire-cut lug brick has advantages not possessed by the repressed brick, and that it is, therefore, better for road purposes. Owing to the keenness of competition, however, the subject is a very delicate one, and is almost always spoken of as a "personal preference." Sometimes one or the other is designated in specifications. That the repressing causes a change in the internal structure of the bricks is shown by photographs of broken bricks printed in a text-book used in the Highway Engineering School of Columbia University (Blanchard and Drowne, p. 554), though points of superiority are not pointed out. It is regularly claimed that the wire-cut lugs furnish a more even and uniform space for filler than can be secured otherwise, and that the bricks made by this process have rougher sides, so that the filler will stick to the bricks to better advantage.

The best brick roads are built on a concrete foundation 4 to 6 inches in thickness, according to the amount of traffic the road is expected to carry. For all ordinary purposes, and especially if the subgrade be fairly solid and uniform, 4 inches of concrete is enough. On this concrete is placed a sand cushion 1½ or 2 inches thick which should be rolled with a hand roller, and on this sand cushion the bricks are placed, with the hardest and best edge upward. The bricks are laid in regular courses crosswise of the road and joints broken as in ordinary brick masonry.

16

Fig. 27.—Brick road in Cuyahoga County, Ohio; 4-inch concrete base and 2-inch sand cushion.

The curb should be of concrete or stone; preferably it should be of concrete and be a part of the concrete foundation built up at the sides so it will be flush with the top of the brick when in place. It is usually the custom in laying the brick to place a board, about $\frac{3}{4}$ inch thick, inside the curb when the bricks are laid.

After the bricks are laid the surface should be swept perfectly clean and the surface rolled with a roller weighing 3 to 5 tons. This is to settle each brick firmly into its bed. Then it is ready for the filler.

There is much difference of opinion among road builders as what kind of filler should be used. Formerly bituminous materials, as pitch, tar, or asphalt, were generally used. These fillers are still used in many sections of the country and their use is strongly defended by those who favor them. On the other hand, the National Paving Brick Manufacturers' Association advocates a cement grout filler, made of 1 part Portland cement and 1 part sand, mixed into a grout. This method also has the support of many competent and experienced road builders.

In applying a bituminous filler the thin board along the curb is removed, and the space filled with the tar or pitch or asphalt, which has been heated in wheeled kettles on the road. Then the hot material is poured slowly into the openings between the courses of brick from a vessel having a spout especially shaped for the purpose. If the filler be what is known as "Coal-tar Paving Pitch," it should be heated to 300 or 350 degrees; if of asphalt the heat should be a trifle greater, perhaps reaching 400 degrees. Often with bituminous fillers two pourings are desirable, one a few minutes after the other, to fill up the spaces to the top of the brick where the material from the first pouring had settled down further into the crevices between the bricks. Care should be taken, in any event, to see that the filling is complete, so that both sides and ends of each brick are perfectly reached by the filler.

Expansion joints must be considered when a cement grout filler is employed. The space occupied by the thin board inside the curb when filled with bituminous material provides for the expansion sidewise, whatever filler be used. In using a cement grout filler it is the general custom to put a similar board crosswise of the road every 25 or 50 feet. Many builders, instead of one board put about four thin boards, say $\frac{1}{4}$ inch thick each, between courses of brick adjoining each other, so as to make a strip about a foot wide across the road to provide for expansion by heat and contraction by cold.

The cement grout filler is mixed in specially prepared boxes, which are placed on legs shorter at one end than the other, so that the grout may be taken out readily with scoops. The mixture, as before stated, should be 1 : 1 of cement and sand, with the proper amount of water to make it run readily. This should be spread by scoops over the surface of the bricks and swept into the spaces between the bricks with paving brooms. This brooming must be done thoroughly, as on the perfect filling of all the joints the excellence of the road depends. All surplus grout must be swept forward, so that the face of the bricks, after the filling has been done, shall be clean.

After the grout has had time to get its first "set," but not giving it time to become hard, the boards along the curb and those crosswise the road are taken out and the spaces filled with a pitch or asphalt filler. If too much time is given, and the grout gets thoroughly hardened, there is likely to be difficulty in getting the boards out.

This form of construction is held by many road builders to furnish the highest type of brick roads. Instances are known where the ground under such a road has been washed out for a distance of 25 or 30 feet, leaving the road standing like a bridge, over which heavy automobiles passed before the underfilling was replaced.

Concrete foundations for brick roads, where there is a reasonably firm subgrade, need not be especially rich. A mixture of 1 : 3 : 6 is usually considered sufficient. That is, 1 part cement, 3 parts sand, and 6 parts gravel or broken stone ranging from $\frac{1}{2}$ inch to $1\frac{1}{2}$ inches in size.

If gravel be used it must be clean, or what is known as "washed" gravel.

Foundations other than concrete have been used extensively for brick surfaces. Old macadam and gravel roads which have been used for years have been leveled down with a scarifier, trimmed to an even grade and surfaced with a road machine, curbs put in, and the sand cushion and bricks put on in the usual way. Sometimes a soft gravel of a self-cementing quality has been successfully used; and in other instances a very soft limestone, such as exists in some sections of the South, has been found to make a satisfactory foundation for brick roads.

Ordinary broken stone or gravel such as is used as the lower course in macadam roads may be used in sections of country where freezing does not extend into the foundations of the road. There are also a number of special foundations, some of them patented, which are worthy of investigation in sections where the materials for concrete or other standard foundations are too expensive.

Building brick roads on sand has become a custom in some sections of the southern part of the country. This custom is perhaps most prevalent in the state of Florida, where the subgrade is usually of a very fine sand, which packs readily. In building these roads the sand foundation is rolled with as heavy a roller as it will carry, usually 6 to 10 tons, and the brick placed directly on the sand. Formerly the curb consisted of planks along the edges held by stakes driven 2 or 3 feet into the sand. In later practice concrete curbs are used, set about 22 or 24 inches deep and coming flush with the brick. Sand is swept into the spaces between the bricks and is called by courtesy a "sand filler."

This type of road cannot be recommended by anyone who has a knowledge of the subject and a fair sense of values. Owing to long freight hauls brick is expensive. The economical use of brick for road surfaces anywhere in the country depends largely on the cost of transportation. In these particular sections the cost of freights alone would pay for a good foundation in some other localities.

While these brick roads "built upon sand" are claimed to cost approximately the same as brick roads built on a good foundation in other localities, the cost of maintenance should practically prohibit their use. The torrential rains to which the far southern sections are subject drives the water through the so-called sand filler, undermines and softens the sand foundation, and permits the surface to become uneven and irregular, and often results in broken-up places in the surface. The annual cost, as a maintenance charge, of taking up the bricks, hand-tamping a new sand foundation, and relaying the bricks to a good surface, is so great that in many cases it would pay several times the interest and sinking fund on the cost of a good foundation and filler.

Brick roads, properly built, as designated by those who have given the subject the most careful study, are among the very best, and, in the long run, economical roads that may be found within freight-rate limits. Improperly built, they are extravagant and unsatisfactory and can serve only a temporary purpose.

Since this chapter was prepared the National Paving Brick Manufacturers' Association has begun the advocacy of the plan of laying brick in concrete mortar on a concrete foundation. To do this successfully the brick should be laid before the concrete foundation becomes thoroughly set. Several problems are involved, and whether they will work out in actual practice remains to be seen.

CHAPTER XVI

CONCRETE ROADS

THE use of cement concrete in the construction of country roads began to attract general attention about 1909. Previous to that time many experiments had been made, and one or two instances had been observed where concrete pavements of twelve to fifteen years' duration had been successful.

The development of concrete roads is shown by the fact that while the amount laid in 1909 would amount to approximately 40 miles of 15-foot roads, the amount laid in 1915 would approximate 215 miles of the same width. (The width of 15 feet is used for convenience. There are 8800 square yards of 15-foot road per mile.)

Probably the greatest and most favorable development of concrete roads in the United States has been in Wayne County, Michigan, the county in which Detroit is located. A considerable mileage has also been built in Milwaukee County, Wisconsin, and in a number of other sections.

There are two classes of concrete roads, each of which has its supporters. They are the "one course" road, where the concrete is mixed and laid to the full depth, and the road finished in one layer of concrete; and the "two course," where one course of a partial thickness of the road is first laid and then followed by a second course to complete the depth, and on which the finished road surface is to be made.

Advocates of a "one course" road claim that owing to the greater mass of material in a solid body cracks are less liable to occur, and that the road generally is stronger and better; advocates of two courses contend that much money may be saved by the use of less cement and a poorer grade of stone in the bottom course than is required in the top course. Probably localities exist where either or both may be correct.

The grading and drainage for concrete roads do not materially differ from those required for other roads, and which are treated in chapters on those subjects.

Forms for concrete roads are very simple. Generally they consist of 2½-inch or 3-inch planks of such width that when set up edgewise they will come flush with the top of the finished road. These planks are held in place by stakes driven deeply and firmly into the ground. Sometimes metal forms are used, and some road builders prefer them as being in the long run more economical and more satisfactory in the work.

The subgrade may either be flat or rounded to correspond with the shape of the finished road. It is generally advisable to finish the subgrade after the forms have been set, as in this manner a greater evenness can be secured with less trouble. When the subgrade is flat, the concrete should be thicker in the center than at the sides, so as to give the surface the necessary crown. However, concrete roads do not require so much crown on the surface as most other types of roads. A rounded slope of 2 to 3 inches from the center to the sides of a 16-foot concrete road will answer all purposes of surface and flood-water run-off.

The mixing of concrete for roads is best done in a batch mixer with a boom attachment which deposits the concrete practically in place on the roadway, where it can be quickly and readily raked and tamped and the surface evened and smoothed properly. Other methods of mixing and placing the concrete are open to serious objections. With hand mixing on mixing boards it is almost impossible to get the concrete mixed evenly and placed on the road quickly enough so as to produce satisfactory results. The same is true of the work of a machine mixer placed at the roadside and delivery made by wheelbarrows. With machine mixers with a revolving tube delivery the fine and coarse material in the concrete have a tendency to become separated while passing through the tube, and more or less of the moisture is likely to drain away. In order to secure satisfactory results with concrete in road building the

Fig. 28.—Depositing concrete on roadway from boom mixer.

Fig. 29.—Finishing the surface of concrete road. Mixer in background.

Fig. 30.—Completed concrete road.

mixing must be done thoroughly and the concrete placed on the road wet enough so that the water will flush to the top under the tamping; and the material must be placed with the greatest regard for the even distribution of the large and small and medium-sized particles in the concrete aggregate; and it should be done within one or two minutes of the time of leaving the mixer. A failure to observe any one of these factors will result in streaks in the concrete, some of which will be weaker than others, and which will be almost certain to develop cracks when the road is completed.

(There are a number of mixing machines on the market with the "boom and bucket delivery." Some of them are so adjusted as to move to new positions by their own power; others require moving by other means. A road builder can readily determine which will best suit his purpose by studying the catalognes of the different machines and writing to road officials where they have been used.)

The material should be a first-class Portland cement, good, clean, sharp sand, and gravel or broken stone ranging from $\frac{1}{4}$ inch to $1\frac{1}{2}$ inches in size. If gravel be used it should be clean of coating or dirt of any kind. What is known as "pit gravel" should be washed artificially, but that taken from streams is usually clean after being screened to size.

The best proportions of material, according to those who have been most successful in building concrete roads, are $1 : 1\frac{1}{2} : 3$; that is, 1 part cement, $1\frac{1}{2}$ parts sand, and 3 parts stone or gravel. Some authorities use a little more sand, say $1\frac{3}{4}$ parts, but the general rule is $1 : 1\frac{1}{2} : 3$. The materials are placed in the mixer and enough water added so as to make a soft mud, and the mixing continues until every particle of stone and sand is coated by the dissolved cement; then the mixer empties it into the bucket, which is run out on the boom and dumped as nearly into place as possible. Then men with iron rakes rake it into place, and tampers follow immediately to pack it down into a close, firm mass. In the raking care is taken, as far as possible in the limited time, to get the larger stones toward the bottom. The tamping must show an even firmness, and if solid material is lacking in any spot a raker should be called back to rake that spot over, and get a better placing of the materials.

The surface, after being tamped, is "struck off" with a template, which usually consists of a board with one edge cut to the proper curve of the road surface. The surface is then usually smoothed with large wooden trowels or "floats"; the men doing this smoothing working from a plank with blocking under the ends resting on the side forms.

During 1916 a new method of finishing was developed. It consists of see-sawing a leather belt across the surface at frequent intervals until the surface becomes hard.

In "one course" concrete the whole mass is laid together and the road practically completed with the finishers but a short distance behind those who are placing the concrete.

In "two course" concrete some of the best road builders use a lower grade of concrete in the bottom course and a weaker mixture. Where stone or gravel suitable for the surface is expensive, the lower course, which is not subjected to wear, may be made of almost any available stone or gravel, and a mixture of 1 : 2 : 4 or 1 : $2\frac{1}{2}$: 4 used. Usually the lower course is mixed and laid the length of the boom on the mixing machine; then the materials for the top course are mixed and laid, and the surface finished as stated.

Expansion joints or contraction joints are insisted on by most road builders, though there are some that consider them unnecessary. These joints are placed 20 to 50 feet apart, according to the judgment of the builder. Usually they are made by putting in a board edgewise, with tar paper on the sides, the board afterward being withdrawn and the space filled with tar or pitch. Some builders use soft iron plates to protect the edges of the concrete, and fill the space between the plates with pitch or tar. There are also some patented joint fillers which have been successfully used. The practice on this branch of the subject is far from uniform and the results variable.

The thickness of a concrete road must be governed by the surrounding conditions. The weight of the traffic, the character of the subgrade, and the climate are important factors. On some heavy traffic roads concrete has been laid to a depth of 7 inches at the center and 5 inches on the sides with satisfactory results. In other cases it has been laid 8 inches and 6 inches respectively, or on a rounded subgrade 7 inches. Two-course concrete is rarely laid less than a total depth of 6 inches, the bottom course being from $\frac{1}{2}$ to 1 inch thicker than the top. If the subgrade be sandy or easily drained, and the climate be such that there is no danger of frost, a thinner concrete may be used than where other conditions prevail. Where the center is thicker than the sides it has been noted that cracks lengthwise of the road are less likely to occur; and where freezing reaches into the subgrade cracks are very likely to appear when the frost leaves the ground, whatever the thickness of the concrete. In some instances roads as thin as 4 inches have given satisfaction. It practically all depends on the local conditions and the care which is given to applying the material to meet those conditions.

Covering the concrete road surface with tarpaulin should follow as soon as the surface is finished and the concrete has its initial set. It is better to wet the tarpaulin so as to retard evaporation from the concrete, which should dry very slowly. After a day or so the tarpaulin should be removed, and the concrete surface covered with earth to a depth of 2 or 3 inches. This earth should be kept sprinkled so as to be moist, but not wet, for a period ranging from a week to ten days, until the concrete has thoroughly seasoned, after which it is swept off. Some road builders admit a certain amount of travel during the last two or three days while the earth covering remains; others do not, and consider it bad practice. When the earth is removed the road may be opened to traffic.

Cracks are very likely to appear in concrete roads, but in many cases and where they are few in number the only damage caused is to the appearance. As soon as cracks appear they should be filled with tar or paving pitch. The cracks are almost invariably due to contraction or to uneven settling; in either case the pitch filling, being compressible, is likely to provide room for the expansion when it recurs.

Old macadam or gravel roads should not be given concrete surfaces if other materials are available. The expansion and contraction of the concrete top, and the impossibility of making a perfect joint between the new concrete and the old base, and with the liability of moisture getting in between, makes such old roads more useful for surfaces of greater flexibility. Concrete is hard and rigid. To avoid cracks and breaks numerous enough to practically destroy the road the greatest care must be taken and every possible condition considered.

Roughening the surface of concrete roads, so as to make better holding for horses, is not advocated by those who have given the subject the most study. The rough surface is said to wear out faster, and it is also held that horses soon get used to a smooth pavement of any kind so that their hauling capacity—considering that the load is on the same smooth surface—is not in any way lessened.

Concrete for roads has its strong advocates and strong objectors. Many miles of good concrete roads have been built, many other miles have been failures. It is claimed, and it is probably true, that those which have failed were not built according to the requirements of the existing conditions.

CHAPTER XVII

BITUMINOUS ROADS

THE general use of bituminous materials in the construction of country roads, particularly in the United States, is of comparatively recent origin. Bituminous pavements, such as sheet asphalt and some other forms, have been used on city streets since about 1870, and in isolated cases some forms of bituminous pavement have been reported at a much earlier date. In some European countries certain "rock asphalts" have been used for many years in city streets.

The development of bituminous roads in this country seems to have resulted from the growth of the experiments with bituminous materials as "dust layers" and "road preservatives." When the automobile travel on the roads began to reach such proportions that the old-time reliable macadam roads were in danger of destruction, hundreds of road officials began to experiment in the hope of finding a remedy. This was about 1906–1907. The first official recognition of the damage claimed to be due to the use of automobiles on the country roads, so far as noted, is found in the Annual

Report of the Massachusetts Highway Department of 1907.

The wide range of these experiments and the publication of their results in scientific and other journals; their presentation at road conventions and in official reports and documents furnished a wide scope for study by those responsible for the upkeep of those heavy traveled roads which were so rapidly going to pieces.

From successful use as dust layers it was found that bituminous materials, properly applied, made good "preservatives." That is, that they could be put on the surface of a road, covered with stone screenings, and that a sort of carpet thus formed would sustain the wear and preserve the road structure indefinitely. (This method is described in Chapter VI, on Road Surfaces.)

From this it was a short step to using bituminous materials in new construction. First, the distribution of tar or asphalt on and into the surface of a new macadam road, so as to act as a binder, was found to work successfully, and called "bituminous macadam." Then the system was elaborated and the stone and bituminous material were heated and mixed before being placed on the road. These were at first also called "bituminous macadam" roads, and were distinguished from the earlier types, the names "penetration method" and "mixing method" being used. Later practice has resulted in further changing the name of the mixing method to "bituminous concrete." (Bituminous macadam—penetration method—is treated in Chapter XIV, on Macadam Roads.)

Development of the mixing method, or bituminous concrete, soon demonstrated that the same principles were involved as in the construction of sheet asphalt pavements. About the only essential differences are that in the bituminous concrete the mineral aggregate, instead of consisting of sand and stone dust, is made of stones ranging up to $\frac{3}{4}$ inch and sometimes larger in size. The heating and mixing was done in the same manner, by machine based on the same principles, only modified so as to handle the coarser material.

In the late 90's Mr. Fred. J. Warren, of Boston, secured a patent on what was named a "bitulithic" pavement, which at once entered into competition for city streets with sheet asphalt. Mr. Warren subsequently alluded to this mixture as "bituminous concrete," it differing principally from ordinary concrete in having a bituminous cement instead of Portland cement as a binder. Later a modification of the bitulithic pavement which was named "warrenite" was prepared for application to country roads.

In England and in Canada refined tars have been used extensively and satisfactorily as the cementing ingredient for bituminous concrete. In other European countries and in the United States its success has not been marked. Within recent years, particularly, most of the practice has been with the use of asphalts of various grades, so that the term "asphaltic concrete" has largely taken the place of the former name.

In 1909 the state of Rhode Island adopted bituminous concrete as a standard type of construction for its state highways. Since that time its use has increased rapidly, and it is now (1917) considered by most road officials to be the highest type of construction applicable to country highways.

Asphaltic concrete can be laid on any good solid foundation. Old macadam and gravel roads when properly leveled down with a scarifier and road machine, and swept clean and imperfect spots replaced, make excellent foundations for asphaltic concrete surfaces. A heavy lower course of new macadam or a well-laid Telford base may be used. But it must be understood and kept in mind that asphaltic concrete is intended, primarily, to supply a durable wearing surface, and only aids the base and foundation by protecting them from injury. It is not supposed to add to the structural or weight-carrying strength of the road, though it does so to some extent by securing the distribution of the weight of the wheel load over more space.

A concrete base 4 to 6 inches in thickness is the ideal base for an asphaltic concrete road. Where the subgrade is firm and well packed 4 inches is ample. Nearly the entire state road system of California, with different surfaces to meet local conditions, has a base of 4-inch concrete. Of course, where the travel consists principally of very heavy loads, such as 10-, 12-, or 15-ton trucks, 5, 6, or even 7 inches of concrete may be necessary. But such cases are exceptional.

The concrete mixture for a foundation or lower course is usually made of a 1 : 3 : 6 mix. That is, 1 part cement, 3 parts sand, and 6 parts broken stone or gravel. No stone in the concrete should be larger than one-half the thickness of the foundation, and in no case should exceed 2½ inches in diameter. Extremely hard stone is not essential, as there is no wear on it; but a fairly tough stone which does not break easily is desirable. In some sections of the country where local stone is of a poor quality a larger proportion of cement may be used, as a 1 : 2½ : 5 or a 1 : 2 : 4 mix. But this is not usually necessary.

In laying the concrete base it is usual to form concrete shoulders at the sides, rising 2, 2½, or 3 inches above the concrete base. These shoulders should be a part of the concrete mass forming the base, and should be of the same height as the asphalt concrete which is to form the surface. For all ordinary road purposes 2 inches are considered sufficient.

Mixing plants, for the mixing of asphaltic concrete,

are a part of the equipment of practically all large road contractors. Sometimes it is considered wise for counties or communities to purchase such plants, especially if they have a large mileage of roads to construct. As a general proposition, however, it is wiser to have asphaltic or other bituminous roads laid by contract. One of the reasons for this is that the preparation and laying of asphaltic concrete requires experienced and capable workman, and practically all of these are in the employ of the responsible road contracting companies. Besides, contractors give a bond for the maintenance of their work for a period of years. Therefore, while it is possible for local officials to purchase equipment and organize a force of experienced asphalt handlers to do the work, it is questionable whether, in actual practice, the results for the money expended will be as satisfactory as though the work be done by contract, and the contractor or contracting company held responsible for the results.

The mixing plants may be what is known as "railroad plants," or the more portable outfits which may be hauled on wheels. The railroad plants are mounted on cars and operated from railroad tracks in railroad yards, or side-tracks, convenient or specially laid for their use. The nearness of the road to be built and the point at which fuel and sand and stone can be received are the determining factors.

Fig. 31.—Portable plant for mixing asphaltic concrete. (Built by Iroquois Iron Works, Buffalo, N. Y.)

In the heater, which forms part of the outfit, the stone and sand, if sand be used, are heated to a temperature of 200 degrees. The asphalt is heated in an-

Fig. 32.—Dumping and spreading asphaltic concrete.

other heater to a temperature ranging from 250 to 350 degrees, according to the particular quality of the asphalt. Under no circumstances must asphalt be heated above 400 degrees.

Fig. 33.—Distributing asphaltic seal coat with machine, instead of by brooming.

The hot stone and sand and asphalt are then fed into the mixing machine in the required proportions, and in a manner which will secure the most complete and thorough results. The mixing continues until every particle of stone, and sand, and even the finest stone dust is

Fig. 34.—Covering the seal coat with sand or screenings.

completely coated with the hot asphalt. Then the mixture is taken to the roadway in specially prepared wagons or wheelbarrows, according to the distance, and deposited on iron dumping-boards, from which it is shoveled to the required place on the road. Work-

Fig. 35.—Completed asphaltic concrete road. (Chevy Chase Road, near Washington, D. C.)

men with iron rakes quickly rake it out so as to make a mass of the right thickness and density.

All this must be done while the material is still hot. The heavy roller follows the rakers as closely as possible, so that the mass becomes compacted while it is still hot, and the rolling continues until the road is hard and firm. A roller weighing at least 10 tons should be used.

A seal coat is then applied. This consists of applying about ½ gallon of hot asphalt per square yard of road, sweeping it over and brooming it into the surface of the asphaltic concrete so that every crevice, or opening, or pore may be filled. On this is spread enough screenings or coarse sand to take up the surplus asphalt. Then the rolling continues until the heavy roller will make no further impression, and the surface is smooth and perfect.

For the best results traffic should be kept off the road for at least twenty-four hours to give the material time to cool. In many cases this is not insisted on; but very heavy loads should not use the road for at least a day. The stone used in asphaltic concrete should be of the best grade obtainable. Even if stone has to be shipped in from a considerable distance, it is of importance that it be hard and tough and with high power of resistance to crushing and to wear.

The size of the stone used in asphaltic concrete varies greatly. Some authorities use stone the largest pieces of which are $1\frac{1}{4}$ or $1\frac{1}{2}$ inches in diameter. Others place a limit of $\frac{3}{4}$ or 1 inch in size. From the largest size the stone is graded down to about $\frac{1}{8}$ inch if sand is used, and to dust if sand is not used. Some road builders use "crusher-run" stone; that is, all the stone from the largest allowable size down which comes from the crusher. This is on the theory that the sizes into which the stone was broken in the crushing process are the sizes that will best work together into a solid mass when mixed with a material which will stick them together. This theory may or may not be correct. In practice it may work out that way sometimes, and not in others.

Some road officials of experience consider the crusher-run plan uncertain and insist on certain percentages of stone of exact sizes, ranging from the largest to the smallest, prepared with scientific accuracy. When this is done there is a large number of authorities to choose from, each having different specifications, and requiring tests of different qualities and sizes to determine the fitness of the stone to perform the functions required of it. There is no question but that the work and finer investigations and analysis of these scientific officials have brought out the methods of determining what is the best practice, and raised the standards of excellence in road building. This fact also applies to the studies in asphalts used for cementing the stone into an asphaltic concrete.

Sand, when used, should range from coarse to fine, and be of a quality known as "sharp" sand; that is, sand consisting of angular grains. In the mixing and rolling each particular particle of stone and sand should be completely coated with asphalt, and should be so compressed that the particles should wedge in between each other and fill up all the spaces, with the asphalt acting as the sticking or gluing substance to hold them together.

The proper selection of asphalt and its proper treatment to form the cement for an asphalt concrete road is of the greatest importance and requires the most careful consideration.

Asphalts are of two general classes: "Native asphalts," or those which are taken from asphalt lakes or natural deposits; and "oil asphalts," sometimes called "residual asphalts," which are the base of some classes of crude petroleum, and which constitute the residuum after the volatile oils, such as naphtha, kerosene, etc., and other by-products have been extracted.

The native asphalts of commerce are the Trinidad asphalt, which is taken from the island of Trinidad, belonging to Great Britain, just off the northeast coast of South America, and the Bermudez Asphalt, which comes from the Republic of Venezuela.

The Trinidad asphalt lake is 114 acres in extent, and is apparently the crater of an old volcano. The surface appears as hard as the surrounding earth, and yet the material is in constant but imperceptible motion. At about 100 feet from land bottom of the asphalt has been found at about 135 feet. In the center of the deposit it has been found impracticable to drill much deeper than that, as the movement of the asphalt bends the tubes and makes further progress impossible. When asphalt is taken from the surface, making excavations 3 or 4 feet deep, the depressions will be found filled in the course of a few weeks. When the material is taken out in large chunks, weighing often 75 to 100 pounds each, and put in the hold of a ship, it is found to have settled into a solid mass by the time it arrives in this country.

Bermudez asphalt is a much softer material. The lake from which it is taken in Venezuela is over 1100 acres in extent and shallow, located in low, flat, marshy ground. When taken out the asphalt is about of the consistency of a stiff jelly. Loaded into a dump car for a few miles' haul to the docks, it settles into a solid mass shaped to the car. Arriving in this country it is removed from the ships by specially prepared dredges.

The native asphalts are refined by being placed in large oblong tanks or vats, heated to a temperature of about 400 degrees, and jets of steam driven through the molten mass until all the impurities are removed, and the asphalt, on being cooled, is ready for the market.

Certain chemical treatment and tests, to produce and determine the exact quality and to insure absolute uniformity, are too technical to be of value in *Practical Road Building*. They belong in the domain of applied chemistry.

Oil asphalts, from petroleums, are of different varieties and classifications. They are generally found in Western and Southwestern oils. The Eastern oils, as those of Pennsylvania and West Virginia, usually have a base of paraffine instead of asphalt. The oils of Oklahoma, Texas, California, and other states in that section, and of Mexico are heavy with asphalt, some of the latter, especially, being claimed to possess so much asphalt that the volatile oils are considered the by-product, and the asphaltic base the predominant one. Such instances, however, if they exist, are exceptions; though almost every oil well presents a different amount of asphalt in its composition.

Whether or not the degree of heat required in distilling the volatile and other products from these oils constitutes a permanent injury to the asphalt residue is a question for the expert chemists to discuss. Their discussions on this and other phases of the subject may be found at great length in various published reports of scientific bodies, technical conventions, and in books printed on the subject. Specifications of various highway officials of scientific leanings may also be found in plenty; with requirements as to specific gravity; flash point (the point at which the material will take fire); ductility; solubility, and a variety of other requirements, most of which differ in the different specifications. Some of these technical specifications are prepared for the purpose of barring out all but one particular brand of asphalt; some will admit more than one, and others will cover almost any material with a fair proportion of asphalt.

Naturally, every producer of asphalt claims a superiority for his product in some particular respect. While his claim may be absolutely true, it is always possible that local conditions and the conditions where the particular material was successful are not the same. The wearing quality of a road, both as to the smoothness and evenness of its surface and the length of time it will last with no repair or slight repair, is the main object in road building; and to the selection of materials, asphaltic and mineral, and to securing the best roads for the money, should the judgment and energy of road officials be directed.

In the judgment of many practical men who are engaged in building roads the local officials who are authorized to conclude contracts should satisfy themselves by correspondence or by personal examination concerning the wearing qualities of the different materials as compared with their cost. Then the specifications under which bids will be received should name the particular materials, especially asphaltic materials, and separate bids called for on each; so that the competition between the different materials should be clearly defined; and allow a consideration of bids in which the cost and guarantees, in connection with the previous records of the materials under similar conditions, may be studied before awarding the contract.

It seems more satisfactory and productive of better results if alternative bids specifying "Bermudez," or "Texaco," or "Aztec," or "Trinidad," or "California," or "Pioneer," or any one of a number of others; or of proprietary cements which do not state the ingredients, should be received, subject to investigation of past performance and cost of construction, with guarantee for a certain period.

In the record of what has been done with the different materials, some asphalts will show a worn-out surface at the end of the bonded guarantee period; others will show surfaces in good condition, and perhaps lasting for years afterward. These matters must be taken into consideration together with the first cost when deciding what material to use and what bid to accept.

Asphaltic concrete roads require little maintenance for several years. Attention should be given them at all times, just as attention must be given all roads under modern conditions of travel. Defective places may appear anywhere, due to faults in construction or to peculiarities in traffic. Generally these may be remedied and the road protected by the application of a thin coat of heavy asphaltic oil and screenings. After a number of years, when the wear becomes great enough to be generally noticeable, an application of hot asphaltic oil or liquid asphalt, about $\frac{1}{2}$ gallon per square yard, covered with $\frac{3}{8}$ or $\frac{1}{2}$ inch of stone chips or screenings, will preserve the road. Reapplications will preserve it indefinitely.

There can be no question but that the asphaltic concrete road, considering the cost, is the highest type of construction for economical country roads except where local conditions may especially favor some other material at less cost.

An asphaltic concrete road, properly built of the proper materials, should not in any climate work into bunches or waves on the surface. Nor should the bituminous material, under the heat of the sun, melt and run to the sides of the road. It may become soft enough in the heat of the sun so that depressions as great as $\frac{1}{8}$ or even $\frac{1}{4}$ inch may be made by heavy loads on steel tires. But if the asphaltic cement be of the proper brand and grade, the surface will restore itself to its original position within a few hours.

With a careful selection of asphalt, based on its previous experiences under similar conditions; and of the stone and sand and other requirements; the authorities having in charge the construction of a first class road must assume the responsibility for the selection of materials and the acceptance of complete bids. (The financial phase of this subject is presented in Chapter IX, on Road Finance.)

It seems appropriate to state, in concluding this chapter on Bituminous Roads, that certain proprietary or patented combinations are alleged to meet certain specifications for asphaltic concrete. If such be the case, it argues an "inside knowledge" of requirement or of performance, or both. All materials and their records should be plainly set forth in any specification for which bids are asked.

It must be remembered that the most valuable properties of asphalt are its adhesiveness, cohesiveness, and its lasting character. A satisfactory asphalt must stick tightly to every particle of stone and sand, and continue to stick after years of wear; it must also hold itself together—not break, or cleave, or crumble with age and service. These are the practical points intended to be covered by technical specifications, but many officials consider it wiser to hold the contractors responsible both for the material and workmanship and the durability of the road as a whole.

CHAPTER XVIII

SAND-ASPHALT ROADS

ROADS made of a mixture of sand and asphaltic materials are a development of efforts begun about 1908 to build a road at reasonable cost on the Cape Cod peninsula. These efforts might be considered as experimental, except for the fact that the officials had a definite idea in mind as to what could be done, and set about to do it.

It required several years to work out the problem. The report of the details may be obtained from the Massachusetts Highway Department, Boston, Mass. They show the progress, from year to year, of compacting the subgrade and applying the asphalt. Then having the road rutted by traffic and injured by other conditions, and smoothed by a drag or a road machine, and applying more sand and asphalt; and finally the development of a good hard road, suitable for use the year round. The information spread, and many localities have availed themselves of it, with the practice modified to meet local conditions.

One of the contractors engaged on the Cape Cod road owned a hotel in a small town in central Florida. He decided to build such a road for a distance of 600 feet in front of the hotel and on some additional drives within the grounds. Equipment and expert workmen were taken to Florida from the contractor's home in

Fig. 36.—Laying sand asphalt near Mt. Doro, Florida.

Connecticut, and the road was constructed according to knowledge gained on the Cape Cod road.

There were differences to be provided for. The Cape Cod sand is mostly rather coarse, Florida sand is very fine. The Cape Cod climate produces hard freezing in winter. In Florida the frost does not enter the ground.

The action of the rays of the sun in the tropical region also had to be considered.

Some years previously a sand-clay road had been built on that 600 feet. The clay of that region is easily

Fig. 37.—Sand-asphalt road near Ocala, Florida. Lime rock shoulders not rolled.

soluble, and the clay had mostly dissolved and the road practically gone to pieces. What there was left of it, however, was ploughed up, leveled, shaped, and rolled hard as a foundation for the asphaltic surface. It is proper to state in this connection that in most sections of the peninsula of Florida the sand which forms the subsoil, and sometimes the soil, will, when moist, pack very hard under a heavy roller; and will retain that firmness for a long time unless exposed to the air or to the washing action of water.

The sand and asphalt were mixed in a portable asphalt mixer, placed on the foundation, and rolled to a depth of about 2½ inches. The sand was heated before mixing. The mixture was approximately 90 per cent. sand and 10 per cent. asphalt, though probably on account of it being a "home job" the exact figures were not kept. The expert workmen in charge used their judgment as to how to make the best and most lasting road.

The road was an entire success; certain large contracting companies shortly afterward began work on considerable stretches of roads, using the same basic principle, modified to suit the requirements of different locations and conditions. Curbs for these roads are usually of concrete and are placed flush with the surface of the sand-asphalt.

While this type of construction is comparatively new and may be considered still in the experimental stage, there appears no reason to doubt that either in the present form or in some development which experience will bring, sand-asphalt roads may be constructed which will answer all necessary road purposes in certain localities for the time being. The cost is low as compared with other types in the sections where they are most available; and in most cases where they have been built or are in contemplation property values do not justify a large expenditure per mile.

Where a community has limited means and a large mileage of roads to improve, for the building up of the interests of the people of a sandy section, sand-asphalt, properly laid, seems to offer the stepping-stone between the sparse valuations and traffic and the heavier demands which will require a more permanent type when the country is better settled, its population and traffic multiplied, and its values ample.

CHAPTER XIX

SPECIAL SURFACE ROADS

THE growth of interest in road building has been such within recent years that many inventions have been made—and some of them patented or copyrighted—which are intended to produce better surfaces or an improvement on existing types of roads, and incidentally make fortunes for their promoters and owners. Some of these have been very successful; others have scarcely been heard from. A few of them are presented here.

Amiesite

This surface is the invention of Dr. Amies, and is controlled by a company having its general offices at Easton, Pa. The peculiarity of this road is that the stone is coated with a bituminous material of such a character and by such methods that it is friable, and may be shipped and hauled and laid cold. The material is usually prepared at a central plant in the vicinity of the stone quarry from which the stone is taken and where it is crushed and screened to the required sizes.

Fig. 38.—Amiesite road (State road) in Bucks County, Pennsylvania.

The lower course is spread on the prepared foundation 3 inches deep and rolled once. It consists of bitumen-coated stone ranging in size from ½ inch to 1½ inches. On this is placed a layer 1 inch thick of similar coated stone of ¼- to ½-inch size and rolled. On the surface of the coated material is spread clean sharp sand to such a thickness as will fill any spaces between the coated stones, and the road is then rolled to a finished surface. The rolling of amiesite should be very thorough to secure the best results, and the roller should weigh at least 10 tons.

Amiesite roads have been laid extensively in New Jersey and Eastern Pennsylvania, and have been generally satisfactory. The foundation of an amiesite road is practically the same as for macadam or bituminous roads, the amiesite material forming the wearing surface.

Burnt Clay

In certain large sections of the Lower Mississippi Valley there are extensive deposits of a sedimentary clay locally known as "gumbo." This gumbo is particularly sticky and plastic when wet, and the roads made of it are practically impassable in wet weather.

The United States Office of Public Roads, Department of Agriculture, worked out a method by which the roads in these sections might be improved. The method consists in placing alternate layers of gumbo and cord wood on the road, after a system of flues have been provided, and then setting the wood on fire. As the wood

Fig. 39.—Preparing to burn clay for a burnt-clay road in Mississippi.

in the four or five layers burns the gumbo is first dried and then burned to a clinker.

As the burning is done on the roadway which has already been graded and ditched, it only remains necessary to use a shallow plow to break up the clinker and mix in enough of the gumbo of the subgrade to thoroughly stick the pieces of clinker together, then smooth off and roll the surface. The result is a smooth, hard road which is said to wear fairly well, and to answer all practical purposes for rural districts.

The construction of burnt-clay roads is necessarily limited to localities where fuel is plenty and cheap. An extended description of the details is published in Farmers' Bulletin 311, Department of Agriculture, Washington, D. C.

Hassam

This type of construction is sometimes known as concrete built by the grouting method. Stones ranging in size from 1½ to 2½ inches are laid on the subgrade and rolled to a depth of 4 inches. A grout, composed of 1 part cement and 3 parts sand, is mixed in a machine mixer and applied through a pipe to the surface of the stone. The rolling continues while the grout is being applied and until all the spaces between the stones are filled.

A top course of 2 inches of smaller sized stone, the hardest available, with trap-rock preferred, is treated in the same manner except that the grout is mixed 1 part cement and 2 parts sand. As a surfacing a thick grout consisting of 1 part cement, 1 part sand, and 1 part pea-stone is swept over and broomed into the surface until all the spaces are filled and the road is smooth. The road should be given several days of "set" before opening to travel.

Rock Asphalt

Rock asphalt is a granulated rock which is impregnated with asphalt, and which, when crushed and placed on a road, can be rolled to a smooth, hard surface. Usually a better result is obtained by heating the asphalt before placing.

Mines of rock asphalt in Germany and Switzerland have been used as pavements in some cities of Europe for many years. In the United States the material is found in Kentucky, Oklahoma, Texas, and California, and to a less extent in some other states. In some localities it has been used quite extensively, and when properly laid has been generally satisfactory.

One difficulty in the use of rock asphalt is found in securing material where the asphalt impregnation is uniform; that is, where the stone carries the same percentage of asphalt. Any considerable variation in this respect is likely to result, especially under heavy traffic, in the wearing of holes in the surface. These are readily repaired by sweeping the dirt out and putting in fresh material heated enough to make it stick and tamping it down with a tamper.

The practical use of rock asphalt is limited to certain

areas in the vicinity of the points of production by the cost of freights.

Some sections of Mexico abound in rock asphalts, some of most excellent grades, and several Mexican cities are mostly paved with it. It has been stated that when Cortez captured the city of Mexico he found the streets paved with asphaltic rock taken from the surrounding hills.

Shells

In several localities along the Atlantic and Gulf coasts of the United States more or less satisfactory roads are made of shells. In Maryland and in the vicinity of New Orleans oyster shells are used, utilizing the raw shells from the great quantities of oysters taken.

Nearly the same method is employed in making a shell road as in a macadam of two courses. The same kind of compacted subgrade is required, and the lower course is about 5 inches, the top course 3 inches. In laying the top course only enough water should be used to sprinkle the surface. As a surface coat a layer of clean sharp sand about $\frac{1}{4}$ inch thick is spread and thoroughly rolled in.

Along some other parts of the coasts are found vast deposits of what may be called prehistoric shell. This is mixed with sand in the natural deposits, and when

Fi . 40.—Shell road in Louisiana.

placed on the road, spread, and rolled to a depth of about 5 inches, makes a good road for horse vehicle traffic. Under the combined traffic of horse-drawn and motor-driven travel it wears rapidly to dust and requires such frequent replacement that the upkeep is expensive.

Any shell road wears into grooves rapidly when the travel follows a certain track. Some road officials in repairing these roads place loose shell in the grooves, which causes the travel to distribute itself over the surface and wear the road more evenly. The state of Maryland has a large mileage of shell roads which are reported as economical and satisfactory.

Other Roads.

Roads have been made by putting about 6 inches of wet wheat straw on an earth road surface, mixing it with the earth with a disk harrow, and rolling. The same method has been followed with sawdust, in the vicinity of lumber mills. With both these materials a fresh application is required once or twice a year.

Slag from blast furnaces when crushed makes an excellent substitute for stone where the furnaces are near enough to justify its use. The slaty refuse from coal-mines after being separated from the coal-dust has been used with more or less success in some localities, where it seems to effect a favorable chemical combination with the soil, producing a fairly hard and durable surface. In other soils the attempt has failed.

Petrolithic roads are made by mixing bituminous material with the natural soil. This is done by ploughing up the earth to a depth of about 6 inches, pulverizing it with a cultivator, sprinkling with water, and applying about 1 gallon of asphaltic oil per square yard. This may be done in one or in two applications. The earth and oil are then thoroughly mixed, the road shaped with a road machine, and rolled with a roller-tamper. In many instances surface coatings of sand and small quantities of asphaltic oil are added in one or two courses, and rolled in with a smooth roller.

Concrete cubes 2 inches square, laid on a foundation of broken stone with a $\frac{1}{4}$-inch sand cushion, have made satisfactory roads. The cubes are made by machinery and allowed to dry for about three months before being placed in the road.

The number of new roads which have been and are constantly being invented precludes the possibility of mentioning them all, much less describing the process by which they are built. Among them are many failures, and undoubtedly some of them will be successful when properly adapted to conditions into which they will fit.

The paramount necessity in the construction of any road, of any material, and by any method, is that good sense and judgment be exercised.

Spotted something that isn't quite right?

Let us know and we'll fix it for future reprints.
And as a thank you for your diligence, we'll
send you a free Sothis book of your choice.

publisher@sothis.press

This page intentionally left blank

Made in the USA
Middletown, DE
05 November 2026

36541616R10180